Philosophies of Structural Safety and Reliability

Philosophies of Structural Safety and Reliability

Vladimir Raizer and Isaac Elishakoff

CRC Press
Taylor & Francis Group
Boca Raton London New York

CRC Press is an imprint of the
Taylor & Francis Group, an **informa** business

First edition published 2022
by CRC Press
6000 Broken Sound Parkway NW, Suite 300, Boca Raton, FL 33487-2742

and by CRC Press
4 Park Square, Milton Park, Abingdon, Oxon, OX14 4RN

CRC Press is an imprint of Taylor & Francis Group, LLC

ISBN: 9781032209302 (hbk)
ISBN: 9781032209340 (pbk)
ISBN: 9781003265993 (ebk)

DOI: 10.1201/9781003265993

Typeset in Times
by Newgen Publishing UK

Contents

Authors

Vladimir Raizer, DSc, PhD has dedicated his entire professional life to strength analysis of civil structures, mainly risk and reliability assessment and probabilistic methods.

During the four decades of his career in Russia, Dr. Raizer served as a fellow in the most prestigious civil engineering establishments in the country – the Central Research Institute of Building Structures, Moscow (1959–1999) – and for 25 years as the head of the Department of Strength and Reliability of Structures. His experience includes analysis of strength, stability, and design on static and dynamic action space structures, domes, reservoirs, pipelines, power transmission lines, masts, probabilistic modeling loads and actions, and processes of failure in multi-element structures, application of the theory of reliability to the problem of risk, safety, and structural survivability, and COD procedures, among others.

Simultaneously, he was also a full professor in the Department of Structural Mechanics, Moscow State University of Civil Engineering (1982–1999). Along with a course of lectures on the problems of strength, stability, and reliability of structures, he led the research of graduate students, of whom more than 30 earned their PhDs.

From 1976–1999, he chaired the Coordinated Council on Structural Safety responsible for developing the National Codes of Practice.

Between 1978 and 1991, he represented his country at the ISO, Technical Committee 98 "Bases for Design of Structures," and chaired ISO TC98, SC3 "Loads, Forces and Other Actions."

Upon immigrating to the US in 1999, Dr. Raizer worked as a consultant at FC&T Corp. in San Diego, CA (1999–2003) and then as an independent consultant providing scientific support for a number of engineering projects.

Dr. Raizer has authored and coauthored more than 150 technical publications and 11 books in Russian and English.

In 1996, Dr. Raizer was appointed as Honorary Scientist of Russia and in 1998 he was elected as Honorary Member of the Russian Academy of Architecture and Construction Science.

Isaac Elishakoff, PhD serves as the Distinguished Research Professor in the Department of Ocean and Mechanical Engineering at Florida Atlantic University. He also holds a courtesy appointment as a Professor in the Department of Mathematical Sciences. He was born in Kutaisi, Republic of Georgia, Europe. Professor Elishakoff holds a master's and Candidate of Sciences Degree (latter being equivalent to the PhD degree in the US) in dynamics and strength of machines from the Power Engineering Institute and Technical University in Moscow, Russia. Prior to joining Florida Atlantic University, he taught 1 year at the Abkhazian University, Sukhumi, Republic of Georgia, and 18 years at the Technion-Israel Institute of Technology in Haifa (1972–1989). He also occupied several visiting positions: an inaugural holder of the Frank M. Freimann Chair Professorship of Aerospace and Mechanical Engineering at the University of Notre Dame, USA; the Henry J. Massman, Jr. Chair Professorship also at the University of Notre Dame; the Castigliano Distinguished Professorship in University of Palermo, Italy; a professorship at the Naval Postgraduate School at Monterey, CA; the inaugural holder of the W.T. Koiter Chair Professorship; a Visiting Professorship at the University of Rome, Italy; a fellow of the Japan Society for Promotion of Science, at the Universities of Tokyo and Kyoto; an Eminent Scholar at the Beijing University of Aeronautics and Astronautics, China; an Eminent Scholar at the Hunan University, China; the Edmond Safra Distinguished Professor at the Technion, Haifa, Israel; a Distinguished Visiting Fellow of the Royal Academy of Engineering at the University of Southampton, UK; and the S.P. Timoshenko Scholar at Stanford University, CA.

Dr. Elishakoff is the author and coauthor of 17 books and 14 edited volumes in addition to 532 papers in leading national and international journals and conference proceedings. His publications have appeared mostly in *AIAA Journal; Journal of Composite Structures; ASME Journal of Applied Mechanics; Proceedings of the Royal Society of London; International Journal of Solids and Structures; Journal of Sound and Vibration; Journal Mathematical Problems in Engineering; Acta Mechanica; Computer Methods in Applied Mechanics and Engineering; Computers and Structures; Journal of Acoustical Society of America; Chaos, Solitons & Fractals; Meccanica; Philosophical Transactions of the Royal Society*, and many others.

In addition to extensive research, he has developed numerous undergraduate and graduate courses, including the first engineering course worldwide, "Design for Homeland Security." Dr. Elishakoff is a recipient of the Bathsheva de Rothschild Prize (1973), as well as has held fellowships from the German Academic Exchange Office, and the National Technical Foundation of the Netherlands. He was presented special medallions of the University of Notre Dame and the University of Tokyo. From 1996–2002, he was appointed as an ASME Distinguished Lecturer. He has been elected as a fellow of American Academy of Mechanics; foreign member of the Georgian National Academy of Sciences; Foreign Fellow of European Academy of Sciences and Arts; fellow of European Academy of Sciences; fellow of ASME; and full member, Georgian Academy of Engineering.

Dr. Elishakoff has lectured at over 200 national and international meetings and seminars, including over 60 invited, plenary, or keynote lectures. His research activities were supported by National Science Foundation (NSF); NASA Kennedy Space Center; NASA Langley Research Center; NASA Glenn Research Center; ICASE-NASA Institute for Computer Applications in Science and Engineering; and National Center for Earthquake Engineering Research. The ASME International Congress and Exposition, held at Lake Buena Vista, FL. in November 2009, organized a special "*Symposium on Stability, Structural Reliability, and Random Vibrations in Honor of Prof. Isaac Elishakoff*". *Special issue of the International Journal of Structural Dynamics and Stability was dedicated to Elishakoff's 65th birthday. A three-volume book, titled Modern Trends in Solid and Structural Mechanics* was edited by Noel Challamel, Julius Kaplunov and Izuru Takewaki in 2021 to mark Elishakoff's 75th birthday. He was awarded Worchester Reed Warner Medal (2016) by the American Society of Mechanical Engineers and the Blaise Pascal Medal by European Academy of Sciences (2021).

Preface

> The business of philosophy is to teach man to live in uncertainty … not to reassure him, but to upset him.
>
> Lev Shestov, *All Things are Possible*

> Any speech which ignores uncertainty is not spoken by a sage.
>
> Ajahn Chach

> Uncertainty is part of the tapestry of life.
>
> Emiljano Citaku, *What If? Your Guide to Making the Best Decisions Ever*

> Science is based on uncertainty.
>
> Lewis Thomas (1980)

The centuries' old experience of engineering activity shows that the problem of safety and reliability of structures has always existed, and it is still relevant today. The concept of safety of designed structures has gone through separate stages and in its main line has always developed under the slogan of more precise prediction of structural behavior, the study of the nature of applied loads, a more distinct description of the requirements to the form of structures, and conditions of execution of such claims.

The history of construction activity shows that even in the most advanced ancient structures, you can find gross errors that reveal ignorance of the basic principles of safety and reliability of structures. Superstitious fear of the unknown mystery of material forced those in construction to seek help from otherworldly forces with the involvement of prayers (which continues even now), conspiracies, and even sacrifices. Since ancient times, the profession of a designer was considered very responsible, and possible construction errors had very serious consequences for those who made them.

Safety standards were usually very vague. It is generally accepted that the most famous written building codes are included in the Code of Hammurabi which dates back to around 1772 BC (Feld, 1964). There are some provisions:

- If a person in construction builds a house for someone and doesn't fabricate it properly, and the house collapses and kills its owner, that person should be executed.

Building codes can be found even in the Bible (Deuteronomy, chapter 22, verse 8): "If you build a new house, then make a railing near your roof so that you do not bring blood on your house when someone falls off it."

Those in construction determined safety intuitively by numerous trials and errors. They learned from the lessons of accidents and collapses of structures. Each accident added new knowledge to the engineers and set new tasks. When the knowledge was not enough, the reliability factor was introduced into engineering design (and is being introduced now). Since no one knew what unpredictable, unidentified phenomena this coefficient considered and whether it should be exactly this, and not less, it was essentially a coefficient of ignorance. After the theoretical foundations for the analysis of structures were formulated in the methods of structural mechanics, it became possible to establish rules for the design of structures and discuss issues of reliability and safety.

Significant and exciting progress is taking place currently in engineering structural design. Difficulties in its complex calculations that seemed to be insurmountable in the past, have practically vanished with the proliferation of "Its Excellency the Computer" in engineering practice. The engineer is now able to undertake more detailed analyses and calculations with allowance for effects and initial data, hitherto excluded or considered approximately.

These new opportunities are now widely utilized to improve the quality of design procedures for ships, aircraft, offshore structures, bridges, masts, pipelines, tall buildings, machinery, mechanisms, and so on.

In parallel, the theory of reliability based on that of probability and on mathematical statistics has progressed in recent decades from abstract scientific research to a means of solution of practical problems in structural design. Its methods are used for codes and procedures, as well as for design of structures with an optimal or prescribed level of reliability.

In this book the authors seek to pool their diverse experience and knowledge. They have some decades of experience in research and analysis of structural reliability in civil engineering. The authors decided to apply the necessary mathematical rigor to the engineering context. They believe that an engineer or a student can master this book even without close familiarity with the theory of probability and mathematical statistics. The authors also believe that this book is about what an engineer cannot afford not to know.

As Doorn and Hansson (2011) stress,

> Decision making regarding safety and risks is philosophically relevant for several reasons. First, different approaches to risk analysis represent different philosophical theories. Standard risk–benefit analysis, for example, is similar to classical utilitarianism in its disregard for persons. Drawbacks of classical utilitarianism pertain to risk–benefit as well.
>
> (Hansson 2007)

In addition, although the establishment of risk exposure limits and other regulations are often presented as "scientific" and "value-free," risk-related decisions often contain value-based judgments on what risks to accept. It typically requires philosophical expertise to uncover these hidden value assumptions in decision making on technological risks (Hansson 2009a).

Third, risk-related decision making requires comparisons of values that are difficult, often seemingly impossible, to compare to one other: loss of human life, disabilities and diseases, loss of animal species, and so on. This issue of value incommensurability is a recurrent problem in philosophy in general and in the philosophy of engineering design in particular.

Thus, you are invited to provide your own feelings and thoughts on the covered topics. We will be most delighted to hear from you, especially by e-mail: vdraizer @hotmail.com or elishako@fau.edu.

Introduction

V. Raizer and I. Elishakoff

PHILOSOPHIES OF STRUCTURAL SAFETY AND RELIABILITY

This book presents three competing philosophies of taking ever encountered uncertainty in structural design in civil, mechanical, aerospace, ocean, marine, and automotive fields. These are probability-based design, fuzzy sets-based analysis, and an approach founded on convex modeling.

Central among these philosophies is the theory of reliability and structural safety and their applications in the probability-based structural analysis, as well as the development of rules, codes, and standards. The book is written to familiarize the reader with the applied techniques of probabilistic design. A probabilistic approach treats all strength, geometric, and deformation characteristics of structures as all actions on them are random values or random processes. An emphasis is given to the basic principles of structural safety and their development. Reliability assessment of manufactured structures is analyzed versus that of representative test specimens and models. Classification of loads and actions, probabilistic models of the main climatic and technological loads on buildings, and their combinations are discussed in detail. Evaluation of acceptable risk and optimization of design parameters in structural engineering are presented. The questions of reliability estimation of the complex systems depending upon their components as well as modern methods of failure probability analysis are considered. Special attention is given to estimation of structural survivability.

Fuzzy sets-based design is illustrated on the alternative approach to safety factors. Authors discuss difficulties that may be associated with probabilistic design, and present yet another alternative to it, namely the convex modeling of uncertainty which is based upon the idea that the involved uncertain variables are bounded. This produces the least favorable response of the structure that can be a design guide.

The book is addressed to specialists involved in the structural design and reliability assessment of a wide range of engineering structures; it is directed to industries, research organizations, regulatory agencies, and undergraduate and graduate students in various fields of engineering.

1 Introduction

> I say that people are unhappy, if everything is clear to them.
>
> Blaise Pascal, *Pensées*

> But to us, probability is the very guide of life.
>
> Joseph Butler

> Life is a school of probability.
>
> Walter Bagehot

The development of the philosophy of safety and reliability of designed structures took place in discrete stages and in its mainstream has always developed under the slogan—more precise prediction of structural behavior, the study of the nature of existing loads, and more clear descriptions of the requirements for the form of structures and conditions for implementation of such requirements.

Any structure is designed to accommodate a certain "technological process"—manufacture and delivery of products and services demanded by human activity. In the course of this process the structure should operate properly during the entire term of its intended service life. Thus, reliability of a structure can be defined as its ability to support normal functioning of the technological process (Raizer, 2009).

Experience has shown that similar structures operating under analogous conditions do not necessarily fail, totally or partially, at the same "age." In fact, it is impossible to precisely assess the actual service life of a structure (i.e., how long it will operate without failure); the best that can be done is to estimate the probability of attaining or exceeding the specific life without failure. ("Failure" can be defined as a situation involving operational trouble or complete arrest of the technological process.) Thus, reliability can be expressed in terms of probability, including both structural safety and serviceability.

Probabilistic interpretations of reliability concepts are commonly accepted in structural engineering. Their fundamentals have been intensively discussed in books, to name a few, by Augusti, Baratta, and Cosciati (1984), Benjamin and Cornell (1970), Bedford and Cooke (2001), Bolotin (1969), Calafiore and Dabbene (2006), Ditlevsen and Madsen (1996), Elishakoff (1983), Faber (2008), Ferry-Borges and Castanheta (1971), Haugen (1968, 1980), Kapur and Lamberson (1977), Kudsis (1985), Madsen, Krenk, and Lind (2006), Melchers and Beck (2018), Murzewski (1963, 1970), Nowak and Collins (2013), Raizer (1986, 1995, 2009, 2010, 2018), Rao (1992), Rzhanitsyn (1971), Smith (1986), Spaethe (1987), Thoft-Cristensen and Baker (1982), and many others, reviewed in (Raizer, 2004). Further advances in probability-based reliability concepts will be focused on probabilistic methods

DOI: 10.1201/9781003265993-1

more suitable for engineering applications, on more effective interpretations of reliability conclusions for real structures, and on assessing the actual uncertainties including those which are not necessarily random.

There are two main ambiguities in using probabilistic methods in design. One is the doubtful possibility of obtaining sufficient raw data for statistical processing. The other stems from inapplicability of the probabilistic approach to unique structures. The answer to the first uncertainty lies in the wealth of modern software packages capable of providing representative probabilistic distributions based on limited data. In regard to the second, it can be said that probability is itself a meaningful measure of the possibility of an event to occur regardless of whether it is recurrent or single. Thus, the probability of failure, usually considered in the theory of reliability, is a characteristic that remains independent of one's approval or disapproval. Interpretation of this probability and study of its genesis are very important.

The sources of random variability of structural and environmental parameters responsible for loads on structures are as follows:

- Statistical uncertainties in the mechanical properties of materials and structures, which exist even when all quality control requirements are satisfied.
- Random character of natural (climatic, seismic, etc.) processes and uncertainty of their onset at the site in question.
- Random human errors and service breakdowns due to negligence, inattention, or incompetence.

All these factors need to be considered independently. The statistics of human errors has been little studied to date; their regularities can only be examined via distant analogies in everyday life such as street accidents.

At the conceptual level it could be important to divide the factors nameable to personnel control in two groups:

- Random fixed factors, that is, random values and random processes with known distributions.
- Uncertain factors with known containing regions or alternatively—if the factor is random—a region with unknown distribution type.

For random factors the term "accessibility" of a random value is an important notion, which is understood as the probability of a random occurrence of this value not being smaller or greater than the design value. This definition makes it possible to derive a formal solution of this problem, but there are interpretational peculiarities with the physical sense of the random values.

For the notion of structural properties, its "accessibility" is associated with an assumption that the structure/material quality meets the pertinent standards. Random processes in the environment (e.g., wind speed, temperature) or in the structure (e.g. corrosion) should be considered in terms of probability of onset of a certain situation during a conditional period of time, fixed at a certain level of confidence. In a distant analogy with life, the case of structural properties would be similar to the probability of infant mortality while the case of random processes in the environment would resemble the probability of death in an accident.

It should be noted that the design values of random natural actions such as snow, wind, temperature, seismic are often defined as those exceeded on average once in T_p years. Then, the onset probability of action F_p in an arbitrary year with the average return period will be $1/T_p$, and the probability P of this action not occurring during the service life T years can be found from the formula:

$$P = (1 - 1/T_p)^T \tag{1.1}$$

All maxima for the period T_p are statistically independent. From Table 1.1 reproduced from AIJ (1996) it follows that this probability is not very high even for large T_p values.

As follows from the table, the probability that during a 50-year period the actual wind speed will exceed its maximum annual value occurs, on average, once in 50 years is rather high and equal to 0.64. However, the reliability of structures capable to bear the load at this wind speed, which is precisely equal to F_p, can be acceptable because the value of load F_p could be sufficiently high.

The probabilistic concept also encompasses the uncertainties in interpretation of the term "limit state." In the case of strength, the limit state is defined as:

$$S_F(F) = S_R(R) \tag{1.2}$$

where $S_F(F)$ and $S_R(R)$ are the load effect and resistance, respectively. As both sides of (1.2) are random values, there is an endless set of these pairs realizing a specific limit state. In the space of the parameters (Figure 1.1) the bisector marks the boundary between the admissible ($S_F<S_R$) and inadmissible ($S_F>S_R$) regions. Since "hitting" the first region (compliance with $S_F<S_R$) can occur only with a certain probability, probabilistic methods are called for analysis of the design situation. By contrast, codes of practice operate with deterministic values of resistance R_p and actions F_p, thus:

$$S_F(F_p) = S_R(R_p) \tag{1.3}$$

TABLE 1.1
Return periods of wind speed vs. 50-year non-exceedance probability of wind load

Return period, T_p (years) for action F_p	10	30	50	100	200	500	1000
Probability of F_p value not being exceeded in $T = 50$ years	0.005	0.13	0.36	0.61	0.78	0.90	0.95

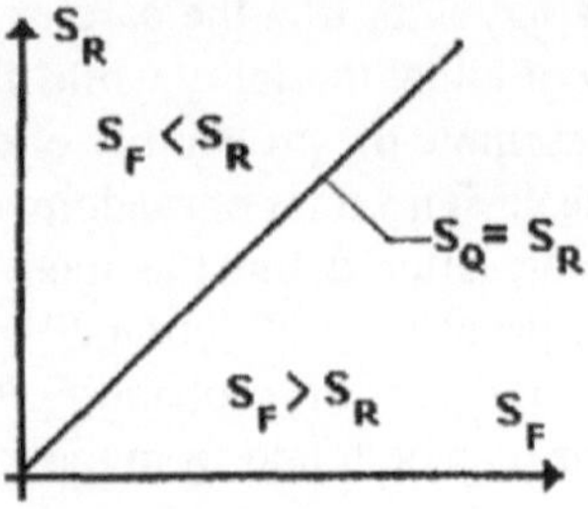

FIGURE 1.1 Space of state and permissible domain

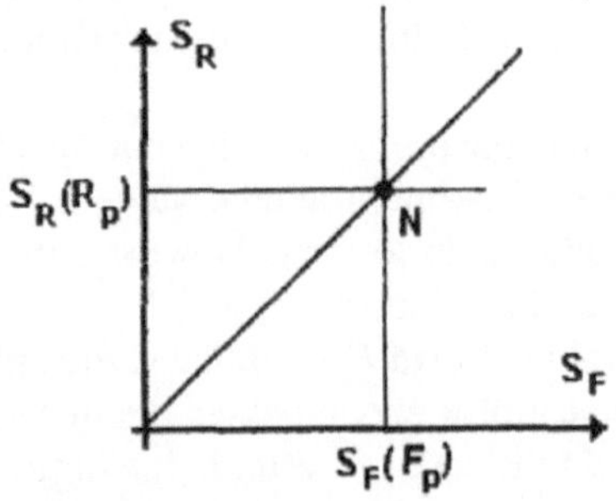

FIGURE 1.2 Codified permissible domain

Only a single point N in Figure 1.2 satisfies the condition (1.3). The admissible state is represented by simultaneous compliance with two conditions:

$$S_F(F) \leq S_F(F_p) \quad \text{and} \quad S_R(R) \geq S_R(R_p) \tag{1.4}$$

In analysis of structures, the reliability level should be estimated using probabilistic models for loads and resistance. In a majority of practical applications, we are interested in rare extreme events with little or no experimental data. In such situations the distributions obtained within a rather wide range of probabilities should be extrapolated to domains with low or very low probabilities. The distributions to be extrapolated are usually those supported by experimental data and information on their theoretical features (asymptotic distributions of extreme values, for example). If the results for low probabilities were received by different numerical simulation techniques such as the Monte Carlo's or some of its modifications, they should be treated with a caution. In such cases modeling should be performed for a theoretical distribution $F(x)$ or for a class of distribution $F(x,\alpha)$ with unknown parameter α, which is supposedly associated with the input data of the problem in question.

While there is nothing wrong with using the Monte Carlo technique in principle, we would like to warn against over-reliance on its results unless they have been discussed additionally. This is important for cases of very low probability, such as 10^{-6} or less. One of the main arguments in favor of determining the reliability by numerical methods is based on the idea that this procedure can be used when no information on reliability function is available. All such calculations can be considered as "experiments" on the structure and should be planned as verification tests for the reliability. For such tests the requirement is

$$N \geq \frac{\ln(1-y)}{\ln P} \tag{1.5}$$

N is the number of no-failure tests with reliability level P and fiduciary probability γ. For highly reliable systems with probability of failure $1-P = 10^{-6}$ and fiduciary probability $\gamma = 0.95$, we shall have $N \geq 3{,}000{,}000$—a situation which raises problems even for a simple structure in numerical analysis. These results can explain, to a certain degree, the well-known fact that experienced engineers are often uncomfortable with the "tails" of distributions and seek an additional safety factor supplementing the theoretical prediction. This precaution is not meaningless.

There is also another problem widely discussed among engineers. A probability of failure equal to 10^{-6}, for example, signifies that one average structure in a million identical ones should be expected to fail. It is hardly possible to find one million absolutely identical structures. These values have no real statistical significance. They represent comparative conditional data, which permit evaluation of different possible design solutions.

This monograph presents the resulting formulations of the theory of reliability and their applications in the probability-based structural analysis.

An emphasis is given to the basic principles of structural safety and the limit state methods in the form of partial factors and their development. Reliability assessment of manufactured structures is analyzed versus that of representative test specimens and models. Classification of loads and actions and their combinations are discussed in detail, as well as evaluation of acceptable and optimal levels of risk in structural engineering. Special attention is given to assessing the robustness in reliability analysis and to applications of reliability theory methods to unique structures and extremely rare and severe disastrous events. The key issues are discussed using specific examples of reliability assessment and probabilistic design of different structures in civil engineering.

The beginning of the 21st century is characterized by an increase in the number of disasters caused by emergencies. You can also add the possibility of terrorist attacks. The book shows how all aspects of the protection of structures that ensure the safety of people, equipment, and structures in case of emergency impacts can be combined into a single aggregated model.

The book is addressed to specialists involved in structural design and reliability assessment of a wide range of engineering structures in civil, marine, mechanical, aircraft & space, and transportation industries, universities, and regulatory agencies.

2 Historical Notes

The safety of the building constructions is a matter of calculating probabilities.

Max Mayer, 1926

The values of safety factors, as well as closely associated values of design loads and design resistances, were improved and modified mainly empirically, by the way of generalization of multi-year experience and exploitation of the structures. Yet, as is seen of the essence of the problem in principle, there are also theoretical approaches possible with wide application of the apparatus of theory of probability and mathematical statistics.

Vladimir Vasilievich Bolotin, 1969

Probability theory provides a more accurate engineering representation of reality. Many leading civil engineers in many countries have written of the statistical nature of loads and of material properties.

C. Allin Cornell, 1969

The times of straightforward structural design, when the structural engineer could afford to be fully ignorant of probabilistic approaches to analysis, are definitely over.

A.M. Lovelace, 1972

As a person who was brought up on factors of safety and used them all his professional life, their simplicity appeals to me. However, if we are to make any progress the bundle has to be unbundled, and each of the constituents correctly modeled.

A.D.S. Carter, 1997

2.1 HISTORY OF SAFETY FACTORS

It is a conventional wisdom to maintain that the scientists and engineers, who earn a living by being engaged in applied mechanics, are divided into two groups: those in the first, traditional group deal with deterministic mechanics, whereas the representatives of the second, more recent, group devote themselves to non-deterministic mechanics. The traditionalists neglect uncertainties in the loading conditions, in the mechanical properties of structures, in boundary conditions, in geometric characteristics, and in other parameters entering into the description of the problem at hand.

DOI: 10.1201/9781003265993-2

The second group embraces itself with various analyses of uncertainty. Within this group, there are those who are active in probabilistic mechanics. They maintain that uncertainty is identifiable with randomness, and hence methods of classical or modern probability theories should be applied. Other investigators developed fuzzy sets-based theories, formulating their analysis on the principle developed by Lotfi Zadeh (1921–2017):

> As the complexity of a system increases, our ability to make precise and yet significant statements about its behavior diminishes until a threshold is reached beyond which precision and significance (or relevance) become almost mutually exclusive characteristics.

The theorists utilizing the fuzzy sets approach base their development on the notion of the membership function, in contrast to the concept of the probability density function utilized by probability proponents. Whereas the two approaches seem to be radically different, they have one thing in common: They both use an uncertainty measure. Engineers developed also an uncertainty analysis that does not demand a measure. The latter method is known by its various appellations: method of accumulation of disturbances, unknown-but-bounded uncertainty, convex modeling of uncertainty, anti-optimization, and, more recently, information-gap methodology.

The set of scientists belonging to the groups dealing with the deterministic mechanics or with the non-deterministic mechanics are not mutually exclusive. A very limited number of scholars simultaneously deal with both approaches. Even smaller numbers of researchers are engaged in seemingly competitive uncertainty analyses techniques. Worldwide, these investigators apparently can be counted on the fingers of one hand. The latter researchers appear to exercise tolerance by holding several contradicting opinions since they find similarities or analogies between the above contradictory approaches, indeed; according to F.S. Fitzgerald,

> The test of a first-rate intelligence is the ability to hold to opposite ideas in the mind at the same time, and still retain the ability to function.

The main goal of this prologue is not a classification of research in applied mechanics, but, rather, to demonstrate that the above characterization, in essence, is imprecise, if not altogether wrong. The main thesis that we would like to propagate is this: In actuality, there is no such thing as deterministic mechanics.

Such a claim may appear, in the superficial reading, to be highly controversial. But once we review the main premises of what is known as deterministic mechanics, we will acknowledge that the above statement is quite transparent. Indeed, although it is assumed within the deterministic paradigm that uncertainty is absent, at the latest stage of the deterministic analysis, after stresses, deformations, and displacements are found by quite sophisticated analytical and/or numerical techniques, somehow, and nearly miraculously, the neglected uncertainties are considered. These uncertainties are enveloped by the concept of SAFETY FACTOR. Thus, the uncertainty is introduced *via the back door*.

According to Vanmarke (1979),

> The format of many existing codes and design specifications impedes rational risk assessment or communication among designers and owners about risk. For example, the conventional safety factor of safety format considers a design acceptable if the computed factor of safety exceeds prescribed allowable value. Such a criterion characterizes structures as either safe or unsafe. It leads many engineers to embracing the concept that all alternative designs which satisfy the criteria are absolutely safe. Consequently, a little thought is given to the ever-present probability of failure, to the factors which influence it, and to opportunities for providing added protection to reduce the risk of failure for better or worse, rigid design provisions such as the factor-of-safety format take much of the responsibility for decision making out of the hands of the engineer. In this sense, codes often serve as cookbooks or as crutches.

According to Ditlevsen (1981),

> the inherent property of engineering quantities remains outside the traditional system of engineering. The system is customarily represented as being deterministic; to each quantity, it is presumed, a unique value can be assigned (e.g., the yield strength of structural steel is 260 N/mn^2, 100 vehicles per hour, make a left turn at the intersection, etc.). In turn structures, pipe networks, earthworks, etc., are conventionally modeled to behave in a unique mechanistic way for a given set of the model's parameter values (dimensions, etc.) and for a given set of input quantities (loads, applied pressures, etc.). The conclusion is that their responses are predictable with certainty.

The inquisitive reader would ask: "What are the reasons embracing the deterministic mechanics and excluding uncertainty?" Ditlevsen (1981) provides the following insight to this possible inquiry: "The reasons for this exclusion have been clear enough: the deterministic representation is easier to learn and use."

Haugen stresses that:

> To design is to formulate a plan for the fulfillment of a human need.
>
> (Haugen quoted in Shigley, 1977)

Krick (1969) considers design a decision-making process: "Initially, the need for design may be well defined; however, the problem can often be somewhat nebulous. There is now a choice of philosophies available for carrying out mechanical design: (1) design based on theory of probability and (2) design based on deterministic assumptions."

This book deals with the following questions: Can uncertainty be introduced *via the front door*? Can the safety factor and the measures characterizing the non-deterministic considerations coexist peacefully? We hope that some partial answers will be provided to these non-trivial questions.

One fact needs to be emphasized. It also spectacularly distinguishes between deterministic and non-deterministic philosophies. While the deterministic method

FIGURE 2.1 The Rambam–Rabbi Moshe Ben Maimon (Moses Maimonides) (born in Cordoba, Spain on March 30, 1135, died in Fostat, Egypt on December 13, 1204; buried in Israel)

claims that once the failure criteria, in conjunction with the safety factor, are satisfied, the structure is immune to failure and the non-deterministic probabilistic approach retains the possibility of failure, unless the probability density is zero outside a finite interval. Thus, the non-deterministic approach appears to be more "honest," putting all consequences on the table, instead of the deterministic approach which discards the failure by relegating it, to a place "under the rug."

Haugen (1980) quotes:

> Since safety factors are not performance-related measures, there is no way by which an engineer can know whether his designs are near optimum or overly conservative. In many instances, this may not be important, but in others, it can be critical.
>
> (Roark, 1965)

Still, we can safely claim that the person who suggested the notion of the safety factor was a genius. This factor allowed and continues to enable constructing safe or nearly safe structures that work. "If it works, use it," one will say. The question is: Could such a methodology be improved? Can we do it better, even though the American proverb advises, "if it ain't broke, don't fix it"?

This monograph maintains that the concept of reliability—the probability that the structure will perform its intended mission—can be used to enhance the deterministic approach based on safety factors. Can we augment the safety factor-based design to remove the mystery from it? The mystery is, of course, in choosing some number, out of the sky, be it 1.2, 1.7, or 2.3 or other that must guarantee the structure's safety? Some justification in choosing the number may also provide a key for the choice of rigorous method(s).

The origins of the concept of safety factor go to times immemorial. You possibly guessed it! The main idea apparently is expressed in the *Torah* (also known as the "Bible"), Deuteronomy 22–8:

> If you build a new house, you should make a fence for your roof, so that you will not place blood in your house if a fallen one falls from it.

This fence from the dangerous edge removes occupants of the house from the possible accident. The Stone edition of the *Torah* (the "Bible") mentions (Sherman, 1993):

> According to Rambam, also known as Maimonides, a Jewish sage and philosopher of the Twelve's century, this commandment applies to any dangerous situation, such as a swimming pool or a tall stairway.
>
> (See Rambam, *Hil. Rotz.* 11: 1–5)

Likewise, engineers put a numerical "fence" below the level of the yield stress. The latter property—that of increase of deformation without increase in the stress level—is easily identifiable with a dangerous situation. Yield stress "warns" us, against eminent danger, and "asks" to install the fence. At this juncture it is instructive to remember the quote by Albert Einstein: "The most incomprehensible thing about the world is that it is comprehensible". This is how, quite possibly, we could speculate about the origins of the safety factor. Materials with the yield stress appear to be perfectly designed to introduce the fence around it. Yield stress, therefore, is both a danger and a blessing: It is easy to recognize that the yield condition is a danger, but it is less trivial to recognize it as a blessing.

One comment is needed here: Moses Maimonides stresses the need to distance oneself from a dangerous situation. In mechanics of solids, we may not have yield stress level alone as an indicator of the dangerous situation, some materials exhibit an ultimate stress instead. Therefore, it makes sense to talk about *resistance* or *capacity* that should not be exceeded by the structure failure criterion.

Hart (1982, p. 118) first poses a question about defining a failure and then replies to it:

> What is *failure*? Failure is what the structural engineer defines it to be and nothing else. For example, of the stress induced by an earthquake exceeds the yield stress of the material, it could be called failure. Alternatively, if the stress exceeds the ultimate stress of the material, it could be called failure. Failure can also be related to structure serviceability ... Therefore, it is fundamentally important to realize that the structural engineer defines failure and that the examples are virtually unlimited.

According to Gnedenko, Belyaev, and Soloviev (1969),

> A *failure* is the partial or total loss or modification of those properties of the units in such a way that their functioning is seriously impeded or completely

FIGURE 2.2 Charles Augustin Coulomb (born on June 14, 1736, in Angouleme, France; died on August 23, 1806 in Paris, France)

> stopped. In certain cases, the concept of failure is sharply defined. A typical example of a component having a well-defined failure is an electric light bulb. The operation of a light bulb has, as a rule, two states: either it gives normal illumination, or it gives no illumination at all. However, in connection with electronic units, the concept of failure is extremely relative since it depends in a significant way on the particular conditions under which the unit may be used.

Petroski (1996) comments about the concept of failure:

> An idea that unifies all of engineering is the concept of failure. From the simplest paper clips to the finest pencil leads to the smoothest operating zippers, inventions are successful only to the extent that their creators properly anticipate how a device can fail to perform as intended. Virtually every calculation that an engineer can perform in the development of computers and airplanes, or telescopes and fax machines, is a failure calculation … What distinguishes the engineer from the technician is largely the ability to formulate and carry out the detailed calculation of forces and deflections, concentrations and flows, voltages, and currents, that are required to test proposed design on paper with regard to failure criterion.

According to Casciati (1991) failure includes.

- "—loss of static equilibrium of the structure, or a part of the structure, considered as a rigid body,
- localized rupture of critical sections of the structure caused by exceeding the ultimate strength (possibly reduced by repeated loading), on the ultimate deformation of the material,

- transformation of the structure into a mechanism,
- general or local instability,
- progressive collapse,
- deformation which affects the efficient use or appearance of structural or non-structural elements,
- excessive vibration, which may cause discomfort and/or alarm,
- local damage (including cracking), which affects the durability of the efficiency of the structure."

Gertsbakh and Kordonsky (1969) classify the reasons for failure as follows:

Construction defects. Failures of this group arise as a consequence of an imperfection in its construction. A typical example is non-consideration of "peak" loads.

A load acting on a system and its elements usually has random variations. In the construction, one tries to keep in mind the possibility of occurrence of "peak" loads, that is, loads considerably exceeding the loads due to normal use. If an analysis and calculation of the loads are made with insufficient care, then the action of "peak" loads will lead to failure. From this point of view of analysis and calculation of reliability, it is important to have the defects in the construction show up to the same extent on all copies of the system or element under consideration.

Technological defects. Failures of this class occur because of violation of the technological manufacturing procedure chosen for the system or unit. The quality of the individual units and connections and of a unit has unavoidable random variations. Quality variations kept within sufficiently restricted limits do not show up appreciably in the reliability of the system. With sharp fluctuations in the quality, the reliability of certain items will

FIGURE 2.3 Claude Louis Marie Henri Navier (born on February 10, 1785, in Dijon, France; died on August 21, 1836, in Paris, France)

FIGURE 2.4 Bernard Forest de Bélidor (born in Catalonia, Spain 1698; died on 8 September 1761, Paris, France)

prove considerably less than the reliability of others. Therefore, technological defects decrease the reliability of some of the items in the total set of manufactured systems or units.

Defects due to improper use. For every system, restrictions are made on the conditions of its use (restrictions on the temperature, on frequency of vibration, etc.) and rules are given for maintenance of the system and its parts, and so forth. Violation of the rules of use leads to premature failures; that is, they increase the speed at which the system ages. Usually, such violations affect only certain used exemplars of the system.

Aging (wear and tear of a system). No matter how good the quality of the unit and the system as a whole, a gradual aging (wear) is inevitable. During the course of use and storage, irreversible changes take place in metals, plastics, and other materials, and the accumulative effect of these changes destroys the strength, coordination, and interaction of the parts and, in the final analysis, causes failures. Thus, variations in the lifetime are caused by variations in the quality of the manufacture, the conditions of use, and aging process.

The earliest possible reference to safety factors in engineering belongs apparently to Bernard Forest who in his book, *The Science of Engineers in the Conduct of Fortification and Civil Architecture*, utilized the safety factor equal to 2.3.

If this book triggers efforts for a clearer justification, augmentation, or replacement of the safety factor by a "better" methodology, its goals will be amply fulfilled.

Indeed, one of the pioneers of the probabilistic analysis of safety factors, Streletsky (1947) notes:

> The concept of the safety factor is directly connected with the security and efficiency of our structures. Despite this fact, one cannot state that this concept was deciphered. Contrary, from all the questions of the analysis of structures the question of the safety factor is most intuitive. Therefore, it is of certain interest to attempt to provide an analytical basis for it.

Since 1947 numerous studies have been conducted to achieve this goal. This book describes some of these developments and, hopefully, provides some novel ideas.

Before we proceed, we ought to reply to a question that a thinking reader may ask: Who was the first investigator who introduced the safety factor? Some investigators ascribe this to Coulomb (Randall, 1977). Bernstein (1957) states that this concept is due to Louis Marie Henri Navier, who also introduced the very concept of stress. He introduced the idea of design according to *working stress* or *allowable stress*, which is obtained by dividing the limiting stress by the factor of safety. Bernstein (1957, p. 49) writes:

> If Galilei was the founder of the science of strength, then Navier was the one who was able to connect it with life: Hence the year 1826—the date of publication of the book by Navier—is not less important in history of this science, the year 1638—the date of its conception.

This is how Bernstein describes the event of publication of Galileo Galilei's (1638) book:

> In 1638 in a private villa in Arcerti outside Florence, where his last days were lived by Galilei, a book was brought from a faraway Dutch city of Leiden. The book was first printed in an Italian language in the publishing firm Elsevier. The book's title was "Dialogues Concerning Two New Sciences." It was necessary to search for a publisher all over Europe, who would agree to publish a new book of the scientist who was condemned by inquisition. Seventy-four years old Galilei took a book, checked it by hands, and put it aside: already a year had passed, since he became blind. In this book of Galilei, the foundations were laid for "two new fields of sciences": dynamics and the theory of strength.

Galilei considered the strength of the beam in 1638; however, it took 188 years, until in 1827 Navier provided a rigorous solution. Readers can consult section 2.6 of Elishakoff's (2004) book for the more detailed account on the priority question (the reader may read short biographies of Coulomb and Navier in the Appendix B).

Prior to completing these preliminary remarks, one has to answer the following nagging questions: Why is the title of the book posed as a dilemma? Why should safety factor and reliability be either friends or foes? Perhaps they are neither, that is, totally unrelated. Perhaps they are both; in other words, maybe they have the "love-hate" relationship encompassing elements of both intricate comradery and enmity. Fischer (1970) in his book on the historians' fallacies describes the so-called "fallacy of false dichotomous questions." He notes, after giving some examples:

> Many of these questions are unsatisfactory in several ways, at once. Some are grossly anachronistic; others encourage simple-minded moralizing. Most of them are shallow. But all are structurally deficient in that they suggest a false dichotomy between two terms that are neither mutually exclusive nor collectively exhaustive.

It appears that a definitive study by Ditlevsen (1988) *Uncertainty and Structural Reliability: Hocus Pocus or Objective Modeling* where the title pinpoints to a dilemma which can be amply defended. It appears to us fully legitimate to raise almost any question, even if it may bear a provocative character. The answer to it may turn out to be that the question itself is not fully valid. Thus, we exercise a more tolerant approach than that adopted by Fischer (1970). For example, the title of the paper by Dresden (1992), "Chaos: A New Scientific Paradigm or Science by Public Relations," appears to these writers quite appropriate: posing questions, even provocative ones, allows for healthy and productive discussion. Indeed, since this book deals with philosophies of structural safety, posing questions appears to be almost mandatory.

As Doorn and Hansson (2011) state,

> The use of safety factors is a well-established method in the various branches of structural engineering. A safety factor is typically intended to protect against a particular integrity-threatening mechanism, and different safety factors can be used against different such mechanisms. Most commonly, a safety factor is defined as the ratio between a measure of the maximum load not leading to failure and a corresponding measure of the applied load. In some cases, it may instead be defined as the ratio between the estimated design life and the actual service life. In addition to safety factors, the related concept of safety margin is used in several contexts. Safety margins are additive rather than multiplicative; typically, a safety margin in structural engineering is then defined as capacity minus load. It is generally agreed in the literature on structural engineering that safety factors are intended to compensate for five major types of sources of failure: (1) Higher loads than those foreseen, (2) Worse properties of the material than foreseen, (3) Imperfect theory of the failure mechanism in question, (4) Possibly unknown failure mechanisms, and (5) Human error (e.g., in design).
>
> (Knoll 1976; Moses 1997)

2.2 DEVELOPMENT OF THE THEORY OF STRUCTURAL RELIABILITY

The theoretical foundation for the design of structures was developed in the middle of the 19th century in the terms of structural mechanics. The first rules of structural design were stipulated in 1840 by the British Chamber of Commerce for railway bridges made of malleable cast iron. Its permissible stress defined as a quarter of the mean limit stress, determined in tests as 20 t/sq.in, and was set at 5 t/sq.in (77.2 MPa) (Pugsley, 1966). This safety factor of 4 was considered acceptable in the first British Building Code for design of buildings with steel framework. The London County Council (Randall, 1977) approved this Code in 1909. The permissible stress for steel was set at 7.5 t/sq.in (115 MPa). Because of unavoidable imperfections, a higher safety factor was proposed for compressed columns. Subsequently the procedure based on permissible stresses was included in the codes of many countries. Mayer (1926) was the first to show the statistical nature of safety factor. In his work he advocated using the probability theory for choosing parameter values. Then Khozialov (1929), considering the randomness of basic parameters, suggested that structural design should be based on optimized capital expenses, probabilities of "defective deviations" and losses through accidents. These two papers were essentially ahead of their time. It was recognized that realistic modeling of structures and loading conditions expected during their working life, as well as the mechanisms of their possible deterioration, should include quantitative consideration of statistical uncertainties of the parameters involved. These ideas were further advanced in the works of Plot (1936), Streletsky (1936), Wierzbicki (1936), and some years later by Freudenthal (1947), Rzanitsyn (1947), and Johnson (1953). These works opened the venue to quantify the safety of structures. Before these works had been adopted in engineering practice, the safety factor as a deterministic value was considered to be a kind of a critical number marking the "failure or no-failure" boundary to be adhered to rigorously. Any, even minor, decrease of the safety factor would lead to jeopardizing the structure, while complying strictly with its value would guarantee a satisfactory level of reliability. Rules of structural design in the codes of practice played an important role in solving the problems of structure reliability. The expected level of reliability is directly correlated to the cost of the structure. A codified level should provide conditions for service without structural deformations, as well as for the necessary durability and survivability. It should also be noted that the necessary level of reliability depends on the design method, type of structural element connection, plan of final tests, and acceptance criteria during production and installation. The theory of reliability had been comprehensively developed in the second half of the 20th century. It covers many kinds of problems. Its origins can be found in the works of notable mathematicians such as Khinchine, Kolmogorov, and Wiener. It is based on the modern theory of probability, the theory of random functions and processes and other mathematical disciplines, and published in a number of books by, to name a few, Bendat and Piersol (1971), Benjamin and Cornell (1970), Gnedenko, Belyaev, and Soloviev (1969), Gumbel (1967), Feller (1970), Parzen (1960), Sveshnikov (1968) and others. The theory of structural reliability was developed mostly independently of the general theory of reliability. Its content and estimation methods have peculiarities, which set it apart from its counterpart, for example, electrical systems. Such peculiarities will be discussed below.

FIGURE 2.5 Letter lambda

2.3 THE CONNECTION BETWEEN ANCIENT SPARTA AND THE FAILURE RATE

Sometimes it is extremely interesting to trace the path of the appearance and development of certain phenomena that seem familiar and studied.

Any reliability specialist is familiar with the Greek letter lambda λ, which indicates the failure rate of a device, such as a diode, capacitor, or electric valve (Figure 2.5). λ is often given in dimension 10 to minus 6 degrees of failures per hour (failures per million hours of operation), and for dimension 10 to minus 9 degrees of failures per hour (failures per billion hours of operation). But where did the lambda itself come from? Obviously, this is the letter of the Greek Alphanumeric alphabet Λ (uppercase) or λ (lowercase), which then turned into the letter L, l.

The most interesting thing was that the shields with the applied lambda were worn by the brave warriors of ancient Sparta—the Greek city-state. Sparta, unlike other Greek polis, with the possible exception of Athens, is widely known in popular culture.

Largely due to the success in military affairs, the courage and bravery of the Spartan soldiers were recognized. It is difficult to meet a person who would not know about the feat of the legendary 300 Spartans.

The lambda could not only be drawn, but also minted. But why not, if we are talking about Sparta? It is simple. Lambda is the initial letter of the historical region of Laconia, the capital of which is still Sparta. And the tough Spartan warriors themselves called themselves the Lacedomonians … Somewhere nearby there is a story about conciseness—the ability to speak briefly and to the point. Here is such a paradox of fate or a game of the forces of chance (or there is no chance in this). Lambda on the shields of warriors who valued resilience, determination, and reliability above all else ("with a shield or on a shield"), it has become an indicator of the reliability of equipment.

3 Safety Factor and Reliability Index

It is remarkable that a science which began with consideration of the games of chance should have become the most important object of human knowledge.

Pierre Simon Laplace (1749–1827)
ThéorieAnalytique des Probabilités, 1812

Statistical thinking will one day be as necessary for efficient citizenship as the ability to read and write.

Samuel S. Wilks, 1951

As PalleThoft-Christensen and Michael J. Baker (1982) note,

> Until fairly recently there has been a tendency for structural engineering to be dominated by deterministic thinking, characterized in design calculations by the use of specified minimum material properties, specified load intensities and by prescribed procedures for computing stresses and deflections.

Uncertainty was introduced by the *back door*, as it were, through safety factors. This chapter discusses connection between structural reliability and safety factors. Deterministic safety factors are replaced by structural reliability based upon statistical and probabilistic philosophy and associated thinking. As Shirley Yvonne Coleman (2013) notes, "The term statistical thinking may have first been coined by Samuel Wilks in his Presidential address to the American Statistical Association" (Wilks, 1951), when he attributed it to the author H.G. Wells, quoting him as saying: "Statistical thinking will one day be as necessary for efficient citizenship as the ability to read and write!" However, according to Tankard (1979), Wilks was paraphrasing H.G. Wells from his book *Mankind in the Making*. For historical accuracy we reproduce herewith the full H.G. Wells quote:

> The great body of physical science, a great deal of the essential fact of financial science, and endless social and political problems are only accessible and only thinkable to those who have had a sound training in mathematical analysis, and the time may not be very remote when it will be understood that for complete initiation as an efficient citizen of one of the new great complex worldwide States that are now developing, it is as necessary to be able to compute, to think in averages and maxima and minima, as it is now to be able to read and write.

DOI: 10.1201/9781003265993-3

3.1 FEATURES OF FAILURE AND PRINCIPLES OF DESIGN

In engineering practice, for many years the safety factor as a deterministic value was considered to be a kind of a critical number marking the "failure or no failure" boundary to be adhered to rigorously. Any, even minor, decrease of the safety factor would lead to jeopardizing the structure, while complying strictly with its value would guarantee a satisfactory level of reliability.

Recall that well-known expression, "is the glass half-full or half-empty?" where for the optimist the glass is half-full, for the pessimist the glass is half-empty: for the engineer, "the safety factor with respect to the volumetric capacity is 2.0." Now, the theory of reliability developed successfully and covered many kinds of problems. It is based on a modern theory of probability, on the theory of random functions and processes and other mathematical disciplines.

The probability of failure is the main numerical measure of structural reliability. Different modes of failure can take place in a structure. So, all possible types of failures should be expected and taken into consideration. Failure constitutes an event, which changes the structure's state from operational to non-operational. One should bear in mind that in complex systems the concept of failure involves a hierarchy of its subsystems and elements. Failures at a low level of this hierarchy can be classified as problems at higher levels but do not necessarily mean loss of working state. For example, the failure of one bolt in a bolted joint means only a defect for the joint as a whole, as long as it continues to bear the load.

The failure of a structure is a random event of realization of one of its damage states. A set of damage states forms a corresponding failure region in a space of parameters of the calculation procedure (a space of states). If the value of the damage is the same throughout the failure region, the realization of the damage state is a clear-cut failure. If the value depends on the depth of penetration into the failure region, then the failure would be a fuzzy failure.

A failure is considered to be a failure-"breaker" when the characteristic for a given type of failure damage occurs at the first realization of the damage state (i.e. instant onset of the damage). If realization of the damage state is followed by a cumulative effect, the failure is considered to be a "disturber." Classification of failures as "breakers" and "disturbers" corresponds to the division of limit states in two groups in the partial factors' method.

In the case of a clear-cut breaker, the probability of complete unfitness of the structure for the use is the only measure of its reliability. In the case of a fuzzy breaker, several damage states of different intensity can be considered, and the realization of each shall formally be considered as a clear-cut failure, probabilities of which are themselves a set of measures of reliability. Such measures could be probability of elastic behavior of the structure, collapse, development of plastic hinges short of a plastic collapse mechanism, and so on. In the case of disturber (both clear-cut and fuzzy) the measure of reliability is the probability of exceeding a maximum applicable value assumed in design, that is, disturber shall also be considered as clear-cut breakers.

The probability of failure of a structure shall be determined as that of an outer action exceeding the load-bearing capacity corresponding to the considered damage state, for example the limit collapse load, limit elastic load, and so on. This probability can be mathematically expressed as a convolution integral where the integrand is a probability density function of the load multiplied by that of the load-bearing capacity of the structure.

Any structures should be designed to operate with a certain level of reliability throughout its construction period and service life. The term "reliability" is understood here as the ability to perform intended functions over an intended period of time. Design calculations usually consider the traditional limit states of structures. The reliability provided at these states ensures that the design values of loads or load effects (e.g., forces, stresses, deformations, displacements, cracks) do not exceed their maximum permissible levels. Two groups limit states are commonly distinguished:

- Limit states associated with complete unfitness for service.
- Limit states associated with restricted normal service.

The first group's states are characterized by collapse of any kind; instability that leads to complete unfitness for service; loss of equilibrium of the entire structure as a rigid body; transformation of a structure into a changeable system; excessive change of form; other phenomena precluding further use of a structure. The second group's states are characterized by achieving the maximum permissible deformations; achieving the maximum permissible level of vibrations; appearance of cracking; crack propagation to the maximum permissible widths or lengths; instability that leads to difficulties in normal operation; other phenomena necessitating restricted use of structure.

Herein there will be considered the influence of factors which determine the stress-strain condition of the structure, peculiarities of components interaction, abnormal behavior, geometrical and physical non-linearity, plastic and time-dependent behavior of materials, appearance of cracks, possible deviation of geometrical parameters from their nominal values, and possible destruction of elements in statically indeterminate systems.

In structural engineering the real structures and actions are replaced by calculation models (schemes) with respect to the considered design situation. Design situations will be evaluated in terms of the consequences of their realization. The following design situations shall be considered (ST 2394, 2015):

- Persistent situations with durations of the same order as the structure's service life and a high occurrence probability (e.g., between major repairs or changes of technological process).
- Transient situations with short durations as compared with the structure's service life, but with a high occurrence probability (e.g., during erection works, major repair, reconstruction).

- Accidental situations with short durations and a low occurrence probability (e.g., resulting from explosions, impacts, equipment accidents, fires).

3.2 SAFETY FACTOR

Safety factor is a deterministic value equal to a quotient where the numerator is the mean value of the resistance (or strength) and the denominator is the load (load effect).

$$z = \frac{\overline{R}}{\overline{F}} \tag{3.1}$$

The following equation can be written for the reliability index β:

$$\beta = \frac{(\overline{R} - \overline{F})}{\sqrt{(s_r^2 + s_f^2)}} \tag{3.2}$$

This term was finally legitimated in document ST 2394 (2015). Earlier, Rzanitsyn (1947) proposed the name "safety characteristics." Snarskis (1962) used the expression "normal distance of failure." In document NKB (1978) the term "safety index" was used.

In the paper by Randall (1977) s_r and s_f are the standard deviations of the random values $\tilde{R}$ and $\tilde{F}$. From Equations (3.1) and (3.2) the relations between safety factor z and reliability index β is

$$\beta = \frac{(z-1)}{\sqrt{\left(v_f^2 + z^2 v_r^2\right)}} \tag{3.3}$$

where $v_f = s_f / \overline{F}$, $v_r = s_r / v_r = s_r / \overline{R}$-variance coefficients. The probability of failure Pf in the case of a normal distribution for R and F will be expressed via reliability index β:

$$P_f = 0.5 - \Phi(\beta) \tag{3.4}$$

where

$$\Phi(\beta) = (1 - \sqrt{2\pi}) \int_0^{\beta} \exp(-x^2 / 2) dx \tag{3.5}$$

The reliability index is the distance between a random value of the so-called strength reserve $\tilde{L} = \tilde{R} - \tilde{F}$ and the midpoint of the distribution. The relation between the probability of no failure (reliability) $P_s = 1 - P_f$ and reliability index β is shown in

TABLE 3.1
Probability of failure vs. reliability index β

β	2.25	3.25	3.75	4.25	4.75	5.25
$P_f = 1 - P_s$	10^{-2}	10^{-3}	10^{-4}	10^{-5}	10^{-6}	10^{-7}

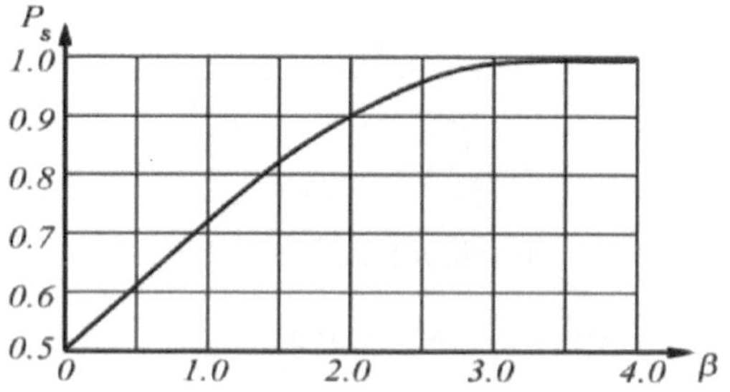

FIGURE 3.1 Dependence of reliability function P_s on reliability index β

Figure 3.1, while the relation between probability of failure P_f and values of reliability index β commonly used in structural design is shown in Table 3.1.

For $\beta \geq 5$ the reliability function P_s is close to unity. Analysis of a large number of structures designed in accordance with different national codes yielded a wide range of variation for β from 1.65 to 10.82 or for the failure probability from 10^{-2} to 10^{-27}.

The quadratic equation (3.3) yields the safety factor z:

$$z = \frac{1 + \sqrt{\left(\beta^2 v_r^2 + \beta^2 v_f^2 - \beta^4 v_r^2 v_f^2\right)}}{\left(1 - \beta^2 v_r^2\right)} \tag{3.6}$$

If $v_f = 0$, then from (1.6) $\quad z = 1/(1 - \beta v_r)$ (3.7)

If $v_r = 0$, then $\quad z = 1 + \beta v_f$ (3.8)

If v_r and v_f are small values, then

$$z \approx \beta\sqrt{(v_r^2 + v_f^2)} \tag{3.9}$$

In most cases there is no or very weak correlation between resistance (or strength) and load effect, and even when present it is usually very difficult to express numerically. A positive correlation $\rho(R, F) = \rho_{rf} \geq 0$ exists when for more solid elements with relatively greater probability a larger load will be expected. This case applies partly to statically indeterminate systems when the higher strength of some elements is associated with their higher rigidity and, as a result, with the higher loads.

A negative correlation between $\tilde{R}$ and $\tilde{F}$ will appear when less solid elements sustain a higher load. Some experiments have shown that under tensile stress, thick-walled elements of steel structures with lightening holes can, as experiments show, sometimes be more rigid than continuous elements due to a 3-D tensile stress in the stress concentration zones.

In cases when the correlation can be evaluated, the formula for reliability index may be expressed as:

$$\beta = \frac{(z-1)}{\sqrt{\left(v_f^2 - 2zv_{rf}^2 + zv_r^2\right)}} \tag{3.10}$$

where $v_{rf} = \sqrt{\dfrac{\rho_{rf}}{RF}}$ and ρ_{rf} are correlation factors.

Then, Equation (2.6) will be transformed to:

$$z = \frac{1 - \beta^2 v_{rf}^2 + \sqrt{\left[\beta^2\left(v_f^2 - 2v_{rf}^2 + v_r^2\right) - \beta^4\left(v_r^2 v_f^2 - v_{rf}^4\right)\right]}}{1 - \beta^2 v_r^2} \tag{3.11}$$

3.3 RELIABILITY INDEX AND PARTIAL FACTORS METHOD

The partial factors method is based on statistical load values, mechanical properties of materials, and state of the structure. The necessary reliability level in this case is determined by codified load values, durability of materials, operational conditions of the structure, and other factors. With the unavoidable ambiguity of structural failure probability, the notion of so-called "assurance" of design values has been introduced, taking into consideration that there is a probability of reaching the limit state at loads less than design values if this is a result of a lesser value of load-bearing capacity (strength).

The design characteristics were determined usually empirically and were based on wide experience in structural design and operation. A theoretical approach can be based on reliability methods by considering jointly random changes of acting loads, materials, mechanical properties, and other items.

The limit state method has two characteristic features. The first is that the limiting states are chosen out of a great number of possible structural states. Also, there are states under which the structure does not meet the service requirements, and the conditions of this non-achievement are marked. The second feature is that all the initial parameters taken as random (having different values with different probability) are given in the codes as determined standard parameters, and the impact of their variability on structural reliability is accounted for in partial reliability factors. Each factor is associated with the variation of only one initial value, thus if the reliability is presented as a multivariable function, then each factor depends on the partial derivative of the function with respect to the relevant variable. This method is deterministic but can be assigned a probabilistic value at any level of reliability depending on the procedures used to determine the partial factors' values. With the limit state method as a means of code development, the structure reliability level is governed only by the choice of design values of loads and resistance for all possible limited states. The procedure of this method can be represented by the following equation,

$$\Psi_F(a_1F_1, a_2F_2, \ldots, a_nF_n) = \Psi_F(bR) \tag{3.12}$$

This equation is structured to correspond to the design values of input parameters, and the latter are determined from the condition:

$$\Psi_F(a_1F_{1d}, a_2F_{2d}, \ldots, a_nF_{nd}) \leq \Psi_R(bR_d) \tag{3.13}$$

In (3.12) and (3.13) F_t and F_{id} are the random and design values, respectively.

R and R_d are those of the resistance and its design value; a_iF ($I = 1, 2, \ldots, n$) and bR are random values of the load effect and bearing capacity; and a_iF_{id} and bR_d are the corresponding design values.

The design values to define the reliability can be chosen for the case when only one load is applied to the structure. In this case the limit state is defined as follows:

$$aF = bR \tag{3.14}$$

and design inequality is

$$aF_d \leq bR_d \tag{3.15}$$

The reliability level is characterized by the probability of non-failure. The design values $P = P\,(aF \leq bR)$ are defined as:

$$\left.\begin{aligned} P_F &= P(F < F_d) = \Phi(\beta_F) \\ P_R &= P(R < R_d) = \Phi(\beta_R) \\ \beta_F &= \Phi^{-1}(P_F): \quad \beta_R = \Phi^{-1}(P_R) \end{aligned}\right\} \tag{3.16}$$

Here, (3.16) characterizes the range of design values of load and resistance, similar to the reliability index (Raizer, 2009).

$\Phi(x) = \dfrac{1}{\sqrt{2\pi}} \displaystyle\int_{-\infty}^{x} \exp(-t^2)dt$ is the Gaussian integral probability distribution.

When equality (3.15) takes place, expression for limit state can be obtained by dividing (3.14) over (3.15):

$$f = r \tag{3.17}$$

where $f = F/F_d$ and $r = R/R_d$ are dimensionless values of load and resistance for which the design values $f_d = r_d = 1$ and the variances and range of design value are

$$v_f = v_F;\ v_r = v_R;\ \beta_r = \beta_R;\ \beta_r = \beta_R \tag{3.18}$$

$$P_s = P(f \leq r) = \int_0^{\infty} P_f(x) p_r(x) dx \tag{3.19}$$

In practical applications, it is desirable to obtain an analytical solution. Since both values are non-negative, this can be achieved using a lognormal probability density function for describing both random variables. When performing qualitative analysis, one can bear in mind that the distribution of density function is of low or any importance. The integral (3.19) can be formulated as follows:

$$P_S = 1 - \Phi\,(\beta) \tag{3.20}$$

where the reliability index is

$$\beta = \frac{\bar{\rho} - \bar{\omega}}{\sqrt{s_\rho^2 + s_\omega^2}} \tag{3.21}$$

Hence

$$\bar{\rho} = \ln \bar{r}; \qquad \bar{\omega} = \ln \bar{f}$$

Parameters s_ω and s_ρ are expressed by numerical characteristics of the distribution as follows:

$$s_w^2 = \ln(1 + v_f^2);\ \ s_\rho^2 = \ln(1 + v_r^2) \tag{3.22}$$

When the variance satisfies inequality $v < 0.4$, we have $\ln\,(1+v^2) \approx v^2$. Then

$$s_w \approx v_f,\ \ s_\rho \approx v_r \tag{3.23}$$

The formula for the reliability index becomes,

$$\beta = \frac{\beta_r v_r + \beta_f v_f}{\sqrt{v_r^2 + v_f^2}} \tag{3.24}$$

$$\text{or}\ \ \beta = \frac{\beta_r \chi + \beta_f}{\sqrt{1 + \chi^2}} \tag{3.25}$$

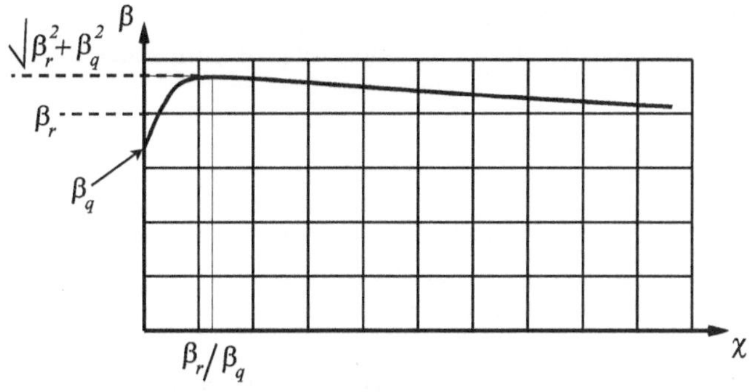

FIGURE 3.2 Dependence of reliability index β on χ-ratio of variation coefficients

where

$$\chi = v_r / v_f \tag{3.26}$$

This means that the reliability index depends not only on the range of indices of design values of loads and resistance, but also on the ratio of its variation coefficients (Figure 3.2).

It will be shown that when deterministic strength value v = $\chi = 0$, then $\beta = \beta_f$, and when deterministic load value $v = \chi = 0$, then $\beta = \beta_r$. Differentiating β over χ and equating the derivative to zero, we have:

$$\frac{d\beta}{d\chi} = \frac{\beta_r(1+\chi^2)-(\beta_r\chi+\beta_f)\chi}{(1+\chi^2)^{3/2}} = 0 \tag{3.27}$$

As $\chi \geq 0$, then $\beta_r - \beta_f\chi = 0$. The maximum probability of no failure will be reached when $\chi = \beta_r / \beta_f$. In this case

$$\beta_{\max} = (\beta_r^2 + \beta_f^2)^{1/2}$$

With the same design values of strength and load, the reliability can vary within a range depending on factor χ. The value of reliability index β and non-failure probability will significantly depend on χ. Returning to the partial factors' method, the general condition of not exceeding the limit states can be presented with reference to (3.13) as:

$$\Psi\,(F_d, R_d, \gamma_D, \gamma_n, C) \geq 0 \tag{3.28}$$

Then for the first group of limit states:

$$\gamma_n \Psi_f(F_d, \gamma_D) \leq \Psi_r(R_d) \tag{3.29}$$

This inequality represents the well-known rule that forces in a structure should not exceed its bearing capacity. For the second group of limit states the inequality can be formulated as:

$$\gamma_n \Psi(F_d, R_d, \gamma_D,) \leq C \quad (3.30)$$

The left-hand part of (2.30) represents the maximum permissible level of displacements, vibrations, crack width or length, or other criteria, while C represents permanent permissible limits of these values. Also:

$F_d = \gamma_f F_{rep}$, where F_{rep} is the representative value of the load; γ_f is the partial factor taking into account possible unfavorable deviations of the loads (on the high or low side) from their representative values due to variability or to deviation in normal usage conditions.

$R_d = R_{rep}/\gamma_m$, where R_{rep} is the representative value of the strength of structural material.

γ_m is the partial factor, considering possible unfavorable deviations of the strength or other properties of material from their representative values.

Possible deviations of an assumed design model from actual working conditions, as well as changes in material properties due to temperature, humidity, duration of an action, its recurrence and other factors not introduced directly in the design shall be taken into account by means of the partial factors of model uncertainties (or working conditions) γ_D. This factor may consider phenomena with acceptable analytical descriptions, such as corrosion due to an aggressive environment and biological influences.

3.4 IMPORTANCE FACTOR

Consequences of financial or social damage resulting from structural failure are accounted for by adding the partial factor of importance γ_n, to the limit inequalities (3.29) and (3.30). It was proposed (Otstavnov et al., 1981, Raizer, 1986) to include an importance factor in the codified procedures. One can easily understand that the consequences of a roof collapse in a kindergarten differ essentially from those in a storage facility. This factor considers the liability of the structure and its influence on the reliability. The required degrees of reliability may be ranged based on classification of entire structures or structural elements. The importance factor γ_n can be selected according to the following failure consequences (Raizer, 2009; ST 2394, 2015):

- Low risk to life; economic, social, and environmental consequences are minor or negligible.
- Medium risk to life; economic, social, or environmental consequences are considerable.

- High risk to life; economic, social, or environmental consequences are highly significant.

Using inequalities (3.13) and (3.25) with $\gamma_n = 1$, there will be variable level of reliability depending on the ratios of variance of the left- and the right-hand parts of equation (3.12) for the limit state. Equation (3.25) supplemented by the importance factor will read:

$$\gamma_n a F_d \leq b R_d \tag{3.31}$$

It can also be mentioned that if all input random values have normal distributions, then Equation (3.24) will retain its form but v_r and v_f will be replaced by s_r and s_f, with $\chi = s_r/s_f$. On the basis of the inequality (3.31), formulas (3.20) and (3.24) will read:

$$Ps = \Phi\left(\frac{\bar{R} - \bar{F} + \ln \gamma_n}{\sqrt{s_r^2 + s_f^2}}\right) \tag{3.32}$$

$$\text{and} \quad \beta = \frac{\beta_r \chi + \beta_f}{\sqrt{1+\chi^2}} + \frac{\ln \gamma_n}{\sqrt{v_r^2 + v_f^2}} \tag{3.33}$$

When $\gamma_n = 1$, (3.33) coincides with (3.25) and the reliability changes within a wide range depending on the ratio of variance of the input values. If $\gamma_n \neq 1$ in the formula for the reliability index, an extra member $\ln \gamma_n / \sqrt{v_r^2 + v_f^2}$ is added and, consequently, β depends not only on the ratio of the variance coefficients, but also on their absolute values. This is shown in Figure 3.3 where reliability index β and

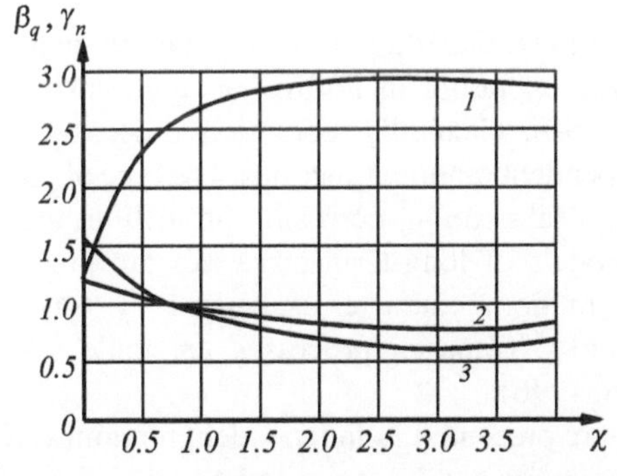

FIGURE 3.3 Reliability index β as a function of ratio $\chi = s_r/s_f$ at $\beta_r = 1.3$ and $\beta_r/\beta_f = 2$ (curve 1) and importance factor γ_n as a function of χ for β_r and β values at $v_f = 0.15$ (curve 2) and $v_f = 0.35$ (curve 3)

TABLE 3.2
Domains of the probability of failure

				$\gamma_n = 1$		$\gamma_n = 0.95$		$\gamma_n = 0.90$	
χ	v_r	v_f	$\sqrt{(v_r^2 + v_f^2}$	β	P_f	β	P_f	β	P_f
0.1	0.05	0.50	0.502	1.90	0.0287	1.80	0.0359	1.69	0.0455
0.2	0.10	0.50	0.510	2.12	0.0170	2.02	0.0217	1.91	0.0281
0.5	0.25	0.50	0.559	2,62	0.0044	2.53	0.0057	2.43	0.0075
1	0.30	0.30	0.424	2.98	0.0014	2.86	0.0021	2.73	0.0032
2	0.30	0.15	0.335	3.02	0.0013	2.87	0.0021	2.71	0.0034
3	0.30	0.10	0.316	2.95	0.0016	2.79	0.0024	2.62	0.0044

importance factor γ are plotted as functions of $\chi = s_r/s_f$. Table 3.2 shows domains for the probability of failure P_f, and reliability indices β for design values of load $\beta_f = 1.65$ and strength $\beta_r = 2.56$ for different γ_n values.

Table 3.2 shows that the level of reliability can differ for structures of equal importance (with importance factor γ_n being the same for the structures in question). Currently the existing rules for structural design do not provide the user any means of assessing in advance whether the structure is designed with the adequate level of reliability. Critical thinking based on understanding the existing design principles allows one to conceptually explore alternative approaches to reliability design, such as optimization procedures or design for a given level of failure probability.

3.5 CONCEPT OF EQUAL RELIABILITY

3.5a Calibration of Model Partial Factor

The partial factor for model uncertainties can be determined comparing the model with similar structures operating in normal or aggressive environment (such as corrosion) (Raizer, 1990). Generally, corrosion processes, for example, can be described by time-dependent random functions. The type of process depends on steel content, the structure's fabrication, corrosion protection, maintenance conditions, and other factors. Models of long-term processes are presented as random time processes, but their random character is caused by time-independent random parameters. Such kind of random processes are called "deterministic random processes" (Middlton, 1996).

When all loads F_i are presented as independent random variables, probability of no failure during service life can be expressed as:

$$P(n) = P[R_1 > F_1,\ R_2 > F_2, \ldots,\ R_n > F_n] \tag{3.34}$$

where R_1, R_2,..., R_n – values of bearing capacity in the considered time intervals. Let's assume $R_n = R_0\varphi(n)$, where $n = t$ is term of maintenance in years; R_0 is the initial random value of bearing capacity; $\varphi(n)$ is a monotonically decreasing non-negative function (i = 1, 2, 3, n) satisfying the following conditions: $\varphi(0) = 1$; $\varphi(\infty) = 0$; $d\varphi / dt < 0$. What should also be mentioned are the additive property of function $\varphi(t)$ and independence of corrosion process at the subsequent time interval t_i of the previous process value at time t_{i-1}, that is, $\varphi(t_1)\varphi(t_2) = \varphi(t_1 + t_2)$. F_1, F_2,..., F_n which is loads in corresponding to considered time intervals.

Let us consider the structure under load F and with resistance R. When random value of the load maximum for a definite period of time (one year, for example) has distribution $P_F(x)$ and the annual load maximums are independent random values, the reliability function can be written as follows:

$$P_1(n) = \int_0^\infty P_F^n(x)dP_R(x) \tag{3.35}$$

It is assumed that there is a structure operating in an aggressive environment and subjected to uniform corrosion. To provide sufficient reliability in the design, additional structural material is necessary for increasing the cross-section area. The condition of no failure is

$$\tilde{F} \le \gamma_D \tilde{R} \tag{3.36}$$

Though the corrosion process is continuous in time, let's consider $\varphi(t)$ as a function of discrete variable n. Condition (3.36) for n^{th} year could be rewritten as

$$F \le \gamma_D R\varphi(n) \tag{3.37}$$

Reliability function will be

$$P_2(n) = \int_0^\infty \prod_{i=1}^{n} P_F\left[\gamma_D x\varphi(i)\right] dP_R(x) \tag{3.38}$$

The equation for defining γ_D can be derived from Equations (3.35) and (3.38).

$$\int_0^\infty \prod_{i=1}^{n} P_F\left[\gamma_D x\varphi(i)\right] dP_R(x) = \int_0^\infty P_F^n(x)dP_R(x) \tag{3.39}$$

For a non-corrosive structural element subjected to axial force, (3.35) can be presented as

$$F \le RA_0 \tag{3.40}$$

where A_0 is the initial cross-sectional area. Function of reliability will be

$$P_1(n) = \int_0^{\infty} P_F(xA_0)dP_R(x) \tag{3.41}$$

For corroding structural element, the cross-sectional area is $A_0\gamma_D$ and $\gamma_D > 1$. The failure condition can be expressed as

$$F \le \gamma_D A_0 \varphi(n) R \tag{3.42}$$

Reliability function (3.38) will be

$$P_2(n) = \int_0^{\infty} \prod_{i=1}^{n} P_F\left[\gamma_D A_0 x \varphi(i)\right] dP_R(x) \tag{3.43}$$

Equality (2.39) for a fixed value of n allows one to determine γ_D. The Fisher-Tippet distribution of maximum value (Gumbel,1967)

$$P(x) = \exp[-(x/\xi)^{-\eta}] \tag{3.44}$$

was chosen for $P_F(x)$. Equality (3.39) yields:

$$\int_0^{\infty} \exp\left[-n\left(\frac{xA_0}{\xi}\right)^{-\eta}\right] dP_R(x) = \int_0^{\infty}\prod_{i=1}^{n} \exp\left[-\left(\frac{\gamma_D A_0 x \varphi(i)}{\xi}\right)^{-\eta}\right] dP_R(x) \tag{3.45}$$

Then it follows

$$\gamma_D = \frac{1}{[n\sum_{i=1}^{n} \varphi^{-\eta}(i)]^{\frac{1}{\eta}}} \tag{3.46}$$

Introducing the corrosion model in the form (Raizer, 2009).

$$P(x)=\exp[-(x/x)-h] \tag{3.47}$$

and expanding the sum in (3.46) in series we can get

$$\gamma_D = \frac{1}{\left\{n\left[\exp\left(\frac{\eta}{\tau}\right)+\exp\left(\frac{2\eta}{\tau}\right)+\ldots+\exp\left(\frac{n\eta}{\tau}\right)\right]\right\}^{\frac{1}{\eta}}} \tag{3.48}$$

TABLE 3.3
Partial model factors γ_D in corrosive medium for different specific number of cavities n and factor η used in Equation (3.48) and introduced in Equation (3.49)

n	η	γ_D in heavily aggressive medium $\tau = 100$	γ_D in moderately aggressive medium $\tau = 150$	γ_D in mildly aggressive medium $\tau = 200$
10	10	1.0666	1.0461	1.0364
	20	1.1304	1.0828	1.0612
	30	1.2098	1.1280	1.0920
15	10	1.0657	1.0439	1.0337
	20	1.1374	1.0851	1.0618
	30	1.2263	1.1354	1.0958
20	10	1.0665	1.0433	1.0327
	20	1.1443	1.0881	1.0632
	30	1.2401	1.1424	1.0999

After transformation we get

$$\gamma_D = \frac{\exp\left(\dfrac{n+1}{\tau}\right) - 1}{n\left[\exp\left(\dfrac{\eta}{\tau}\right) - 1\right]} \tag{3.49}$$

The model factors obtained in accordance with (3.49) are given in Table 3.3.

Aggressiveness of environment can be classified depending on parameter τ: heavily aggressive $\tau = 100$, moderate aggressive $\tau = 150$, mildly aggressive $\tau = 200$.

3.5b Reliability of Transmission Lines

This approach is illustrated by an analysis of transmission lines. Reliability of transmission lines depends on their extension. The longer the line, the greater is the occurrence probability of a wind speed capable of destroying its supports and causing failure of the transmission line. When designing the transmission line this situation should be considered with a choice of the design value of wind speed. If equal reliability is required for transmission lines of different extents, then the line with a longer extent should be designed to a higher design wind speed. The following assumptions are made:

- Supports are subjected to wind loads only.
- Dynamic effect of wind actions is ignored.

- Maximum wind speed value w at the line's span of length 1 at time t_i is assumed to be the Fisher-Tippet 1 distribution with cumulative probability function (CDF) (Gumbel, 1967).

$$P(x) = \exp\left[-\exp\left(\frac{a-x}{b}\right)\right] \tag{3.50}$$

- Over different parts of the length "*l*" the maximum values of wind speed are independent realizations of the same random value in time t_i.
- Resistance of the line's support "*x*" is expressed in the same units as wind speed *w* and is presented as an independent realization of the same random value with density function $p_w(x)$. Gaussian probability density function is accepted.

$$p_w(x) = \frac{1}{s_w(x)\sqrt{2\pi}}\exp\left[-\frac{(x-\overline{x})^2}{2s_w^2(x)}\right] \tag{3.51}$$

Let's consider two transmission lines, the first of length *l* and designed to the wind action w_1, the second of length *ml*, and designed to the wind action w_2. The reliability function of the first line can be presented as follows:

$$P(t) = \int_0^\infty p_{w1}(x)\left[P_{w1}(x)\right]^{t/t_i} dx \tag{3.52}$$

where *t* is service life of the structure.

Taking $w_2 = nw_1$, the transformation formula of random values gives

$$p_{w2}(x) = \frac{1}{n}p_{w1}\left(\frac{x}{n}\right) \tag{3.53}$$

Substituting (3.50) and (3.51) in (3.52) we may write for the first line

$$P(t) = \int_0^\infty \exp\left[-\exp\left(\frac{a-x+b\ln(t/t_i)}{b}\right)\right]\frac{1}{s(x)\sqrt{2\pi}}\exp\left[-\frac{(x-\overline{x})^2}{2s^2(x)}\right]dx \tag{3.54}$$

The function of reliability for the second line can be written similarly

$$P_m(t) = \int_0^\infty \exp\left[-\exp\left(\frac{a-x+b\ln(mt/t_1)}{b}\right)\right]\frac{1}{ns(x)\sqrt{2\pi}}\exp\left[-\frac{(x/n-\overline{x})^2}{2s^2(x)}\right]dx \tag{3.55}$$

To provide equal reliability of transmission lines with different lengths, Equations (3.54) and (3.55) should be the same at each *t*. The problem consists in finding such value n at which these functions would be close to each other. An approximate evaluation was obtained from the condition:

$$\left.\frac{dP}{dt}\right|_{t=0} = \left.\frac{dP_m}{dt}\right|_{t=0} \tag{3.56}$$

Condition (3.56) with (3.54) and (3.55) becomes,

$$G_0 n^2 + G_1 n + G_2 = 0 \tag{3.57}$$

where

$$G_0 = \frac{s^2(x)}{2b^2}; \quad G_1 = -\frac{\overline{x}}{b}; \quad G_2 = \frac{\overline{x}}{b} - \frac{s^2(x)}{2b^2} + \ln m.$$

Let's replace $b,\ \overline{x},\ s^2(x)$ with dimensionless variation coefficients of support's resistance $v_x = \dfrac{s(x)}{\overline{x}}$
and maximum wind speed

$$v_w = \frac{1.283b}{a + 0.5776b} \text{ and } k = \frac{\overline{x}}{a}$$

Using the known equality (Raizer, 1986)

$$\frac{a}{b} = \frac{\pi}{v_w\sqrt{6}} - \gamma$$

where $\gamma = 0.577 -$ Euler constant, equations for $G_0,\ G_1,\ G_2$ can be rewritten as follows:

$$G_0 = \frac{1}{2}\left[v_x v_w\left(\frac{\pi}{v_w\sqrt{6}} - \gamma\right)\right]^2 \quad G_1 = -k\left(\frac{\pi}{v_w\sqrt{6}} - \gamma\right)$$

$$G_2 = k\left(\frac{\pi}{v_w\sqrt{6}} - \gamma\right) - \frac{1}{2}\left[v_x k\left(\frac{\pi}{v_w\sqrt{6}} - \gamma\right)\right]^2 + \ln m \tag{3.58}$$

For $v_x = 0.125$; $k = 3$ the results of calculations are presented in Table 3.4. Factor n in the last column shows how much the design wind speed should be increased if the line's length is doubled or tripled. One can see if, for example, the line's length is doubled, its reliability can remain unchanged by increasing the design wind speed by 2.7%. For the case of $v_w = 0.2$ (first line in the Table 3.4), probabilities of no failure $P(t/t_i)$, with t/t_i being a dimensionless number of years, are given in

TABLE 3.4
Design wind speed increase

v_w	*m*	*n*
0.2	2	1.055
0.2	10	1.235
0.3	2	1.075
0.3	10	1.258

TABLE 3.5
Probabilities of no failure

t/t_i	$M = 1$	$M = 2$
5	0.99955635	0.99955842
10	0.99911097	0.99912128
15	0.99867585	0.99869675
20	0.99824924	0.99828243

Table 3.5. The probabilities are almost the same at different service lives, which confirms validity of the proposed assumptions.

3.6 DEVELOPMENT OF RELIABILITY-BASED DESIGN APPROACH

For the existing codes, the main design requirement for structures is formulated as the inequality $R_d \geq F_d$, where R_d and F_d are design values of carrying capacity (or resistance) and load effect. Realization of this inequality is considered to be sufficient for assurance of the proper level of reliability. In actuality, this level can vary in a wide range and can be insufficient in some situations. In such cases the structure's reliability level has to be assessed.

The basic design requirement can be formulated as

$$P_{rob}(R < F) = P_{fi}^{ex} \tag{3.59}$$

where R and F are the random values of resistance and load effects, P_{rob} is the probability of failure, and P^{ex} is the expedient value of this probability of the *i-th* type failure. The Equation (2.59) is the mathematical form of customer requirement to the structure. The codified reliability requirements can be written in general as:

$$P_{fi}(T) = P_{rob}[g_i(r_1, r_2, \ldots, r_n) < \Delta_i] - P_{fi}^{ex}(T) \tag{3.60}$$

where g_i is the function of working capacity in the *i-th* type of failure where $i = 1,2, \ldots, m$; m is the number of failures; r_j is the random parameter including loads, forces, resistance and so on, with $j = 1, \ldots, n$; Δ_l is the border of failure; T is the time period; P_{fi} is the probability of the *i-th* type failure; P_{fi}^{ex} is the expedient value of the probability of the *i-th* type failure. If the index of the codified reliability is β, then the codified reliability requirement can be written as:

$$\beta_i = \beta_i^{ex} \tag{3.61}$$

where

$$\beta_i = \Phi^{-1}[1 - P_{fi}(T)], \quad \beta_i^{ex} = \Phi^{-1}[1 - P_{fi}^{ex}(T)]$$

and

$$\Phi = \frac{1}{\sqrt{2\pi}} \int_0^x e^{-0.5\xi^2} d\xi$$

In the existing codes (ISO, Eurocode1, USA and Canadian building codes) the requirement of comparing the design load with the strength, as, for example, in formulas (2.16), should be replaced with the requirement of comparing the design probability of failure with the codified rational value of this probability, as in formulas (2.59) and (2.60). A new concept of codification and standardization for evaluation the probability-based reliability of structures should be consistent with the described below regulations. These regulations can be considered an independent document.

- Reliability of structures is their ability to perform intended functions during the intended time. Numerically, this ability is characterized by the measures of reliability.
- Structure reliability can be measured as the occurrence probability of state of damage which is a state that predetermines a considerable economical damage. The values of reliability measures should be calculated by means of probabilistic methods, which consider the random nature of inner properties of structures and external actions on them.
- Design models should reflect actual structures and actions. Uncertainties, which are not covered by design models and calculation procedures, may be considered when formulating the failure conditions. Also, random consequences of failure should be considered when assuming the necessary level of analyzed probability. Total realizations of all calculation parameters (internal properties of the structure and external actions on it) determine the state of the structure.
- Failure of the structure is a random event of implementation of one of its damage states.

- A set of damage states forms a failure region in the space of calculation parameters (the space of states).

A clear-cut failure should constitute the implementation of one of the damage states, while the occurrence probabilities of the damage state should be considered as a set of measures of reliability. Such measures could include probability of elastic behavior of structure, probability of collapse, probability of development of plastic hinges, and so on.

- Probabilistic models for all basic variables should assume that these variables are random and are characterized by their probability distribution functions. Such random variables for time-dependent processes are their maximums for the period of time when the reliability of the structure is evaluated.
- When dealing with several time-dependent loads, their probabilistic models should be based on products of random variables factored by combinations factor accepted in design.
- Failure probability of structure should be determined by the probability of an external action exceeding load-bearing capacity to the considered damage state, such as limit collapse load, limit elastic load, and so on. The failure probability can be mathematically expressed as a convolution integral whose integrand is the probability density function of load multiplied by a probability distribution function of load-bearing capacity of the structure.
- When calculating the probability distribution function of load-bearing capacity of structure, the pattern of variability of numbers that determine strength of its elements should be considered. Thus, if probability distribution function of numbers determining strength of material is developed on the basis of test results, obtained for the specimens selected from the material population available for fabrication of the given structure, then the numbers determining the strength of the elements might be considered independent of each other. If material tests did not take place but material numbers determining the strength represent the population produced in the country, then the correlation between strength quantities of the elements of the structure should be considered. When no information of this correlation is available, one can accept double-sided bounds to the measure of reliability assuming the independence (or functional dependence) of elements' strength quantities of each other.
- Taking several independent loads into consideration, the failure probability of structure should be formulated as a multiple convolution integral. In this case, successful integration with respect to each variable should be performed.

4 Evaluation of Failure Probability

When we work for tomorrow, and do so on an uncertainty, we act reasonably; for we ought to work for an uncertainty according to the doctrine of chance which has been demonstrated.

(Blaise Pascal, 1623–1662)

The probability of failure doesn't matter to me.

(Vinod Khosla)

The main problem of probability-based design is in calculating the failure probability. It is very important to notice that the purpose of calculations is not a number but an understanding.

(Hamming, 1987).

4.1 GENERAL COMMENTS

The condition of failure is mathematically expressed by the inequality R-F<0, where R is the resistance or load-bearing capacity and F is load effect, both considered as random values. In some cases, considering the rigidity requirements (serviceability limit state), R can be the definite limit. Thus, the failure probability is the probability of the inequality realization

$$P_f = P_{rob}\left(R - F < 0\right) = \int_0^{\infty} P_R(x) p_F(x) dx \tag{4.1}$$

where P_f is the probability of failure; P_R is the probability distribution of quantity R; and $p_F(x)$ is the probability density function of quantity F. Integral (4.1) can be also presented in the form:

$$P_f = P_{rob}\left[g\left(R,F\right) < 0\right] = \iint_{\Omega} P_R(x_1,x_2) P_f(x_1,x_2) dx_1 dx_2 \tag{4.2}$$

In this case g is the performance function and Ω is the failure region. The values of R and F can be considered as functions of many variables such as load effects,

DOI: 10.1201/9781003265993-4

characteristics of strength, rigidity, geometrical parameters, and so on. So, integral (4.2) can read:

$$P_{rob}\left[g\left(x_1,x_2,...,x_n\right)<0\right]=\iint\limits_{\Omega_n} P_R(x_1,x_2,...,x_n)P_F(x_1,x_2,...,x_n)dx_1dx_2...dx_n \qquad (4.3)$$

where Ω_n is the failure region in the n-dimensional space. In some cases, integral (4.3) can be written as:

$$P_f = P_{rob}(g<0) = \int_{-\infty}^{0} p_g(\xi)d\xi \qquad (4.4)$$

where $p_g(\xi)$ is the probability density function of quantity g.

The methods of determining the failure probability are the methods of calculating integrals in Equations (4.1)–(4.4). These procedures have some peculiarities. If a distribution of initial values is different from the normal law and there is a nonlinear boundary of the failure region, the probability of failure is calculated by means of the method formulated conceptually in (Hasofer & Lind, 1974). The authors called it "First Order Reliability Method" (FORM). Snarskis (1972) proposed the same method under the name "Hot Point Method." The accuracy of the Hasofer-Lind method was investigated by Elishakoff and Hasofer (1985) and Elishakoff (2004).The work by Rackvitz and Fiessler (1978) ought to be mentioned as well. Later, some authors used this approximate method with different modifications as FOSM, SORM (First Order Second Moment Method and Second Order Reliability Method, respectively) (Raizer, 2004). A technique presented here is based on local approximation of basic variables distributed normally.

Besides that, other numerical methods are widely used, such as direct integration of distribution function and mostly the Monte-Carlo method. As expressed by the leading founder of the modern field of statistics Karl Pearson (1857–1936), "The record of a month's roulette playing at Monte-Carlo can afford us material for discussing the foundations of knowledge."

These methods are illustrated using the following example. Eccentrically loaded compressed steel column with I-section (Figure 4.1) is considered. Geometric characteristics of its cross section are moment of inertia $I = 46.12$ in^4 (1920 sm^4), section modulus $W = 14.64$ in^3 (240 sm^3), sectional area $A = 7.44$ in^2 (48 sm^2). Eccentricity of the normal force N, $e = 0.5$ in (1.2 sm). Critical force $N_e = \pi^2 EI / l^2 = 1.2 \times 10^3$ kNE – modulus of elasticity, l – bar's length.

The probability of exceeding the yield stress in any fiber of a cross section is taken as a failure criterion. The limit state function $g = g\ (x_1, x_2, ..., x_n)$ can be written as follows:

$$g = \sigma_y - (Ne/W)\eta - N/A = N[(e\eta)/W - 1/A] \qquad (4.5)$$

where $\eta = 1/(1 - N/N_e)$; σ_y is the yield stress. In the case of failure $g<0$, and $g = 0$ represents a boundary of the failure region.

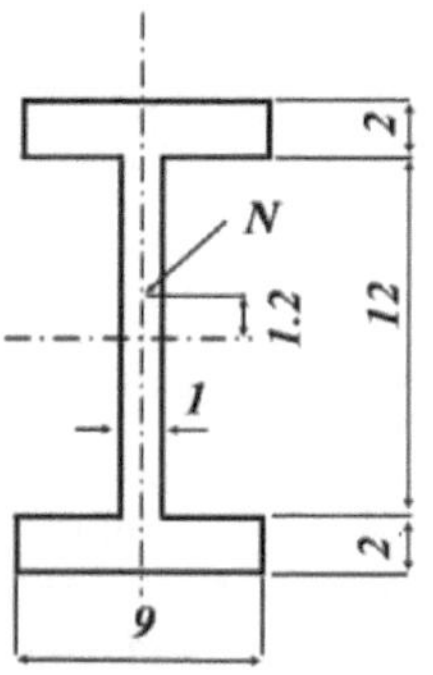

FIGURE 4.1 Cross section of the column

The load is assumed to be distributed according to Fisher-Tippet Extreme 1 distribution with CDF for the normal force N:

$$P_N(x) = \exp\{\exp[(x-a)b]\} \tag{4.6}$$

Probability density function is:

$$p_N(x) = (1/b)\exp[-(x-a)/b]\exp\{-\exp[-(x-a)/b]\} \tag{4.7}$$

Parameters of this distribution are $a = 5 \times 10^5$ H, $b = 6 \times 10^4$ H. These yield:

$$\bar{N} = a + 0.577b; \quad s_N^2 = 1.654b^2 \tag{4.8}$$

Relations 4.8 are reduced to: $\bar{N} = 5.35 \times 10^5$ H, $s_N^2 = 7.69 \times 10^4$ H. Yield stress σ_y has lognormal density

$$p_{\sigma_y}(x) = \frac{1}{xt\sqrt{2\pi}}\exp\left(-\frac{\ln x - m}{2t^2}\right) \tag{4.9}$$

Parameters of this distribution can be taken from the equations:

$$m = \ln\bar{\sigma}_y - 0.5t^2; \quad t^2 = \ln\left[1 - \left(\frac{s_{\sigma_y}}{\bar{\sigma}_y}\right)^2\right] \tag{4.10}$$

Numerical characteristics are $\bar{\sigma}_y = 3 \times 10^8$ Pa, $s_{\sigma_y} = 3 \times 10^7$ Pa, and from (4.10): $t = 0.1$, $m = 19.51$. With these values and denoting $x_1 = N$, $x_2 = \sigma_y$, Equation (4.5) will transform to

$$g = x_2 - x_1[60 \times 10^6/(1.2 \times 10^6 - x_1) + 208.3] \tag{4.11}$$

That yields:

$$\partial g/\partial x_1 = -7.2 \times 10^{12}/(1.2 \times 10^6 - x_1) - 208.3,$$

$$\partial g/\partial x_2 = 1 \tag{4.12}$$

4.2 "HOT POINT" METHOD

The "hot point" method is based on local approximation around a point of the failure region boundary with maximum joint probability density function of all basic variables. If the failure region boundary is nonlinear, then its linearization is performed at this point. The space is transformed to a standardized normal one where the "hot point" is the nearest to the origin of coordinates among all other points of the failure region boundary. If the quantities R and F are normally distributed random variables, the integral in (3.1) should be expressed via standardized normal distribution function Φ:

$$P_f = 1 - \Phi(\beta) \tag{4.13}$$

where

$$\beta = \frac{\bar{R} - \bar{F}}{\sqrt{s_R^2 + s_F^2}}$$

is the reliability index, $\bar{R}$ and $\bar{F}$ are the mean values of quantities R and F; s_R and s_F are the standard deviations of quantities R and F and β is the distance from the origin of coordinates to the "hot point." The condition for choosing this point ensures the minimal error margin of the method.

Searching the "hot point" is an iterative process. At the beginning, one should choose the point by approximation and linearization (the adjustment point). Then, the location of the "hot point" should be defined as the nearest point to the origin at the failure region boundary. If these points do not coincide, then a new location of the "adjustment point" should be sought. The process continues until the adjustment point turns out to be the "hot point." The diagram of the method is shown in the Figure 4.2.

During the first iteration the adjustment point can be chosen arbitrarily at the boundary $g = 0$. Coordinates of the adjustment point can be taken, for example, as the fractals of probability distribution equal to 0.95 for all initial values but one only. The latter coordinate can obtained from condition $g = 0$. If there are initial random values $(x_1, x_2, \ldots, x_n)$ with well-known cumulative F_{xi} and density f_{xi} for the probability distribution functions, then coordinates of the adjustment point can be determined by the formula

$$x_i^{ad} = F_{x_i}^{-1}(0{,}95)\Big|_{i=1}^{n-1} \tag{4.14}$$

and x_n^{ad} can be found from condition $g = 0$.

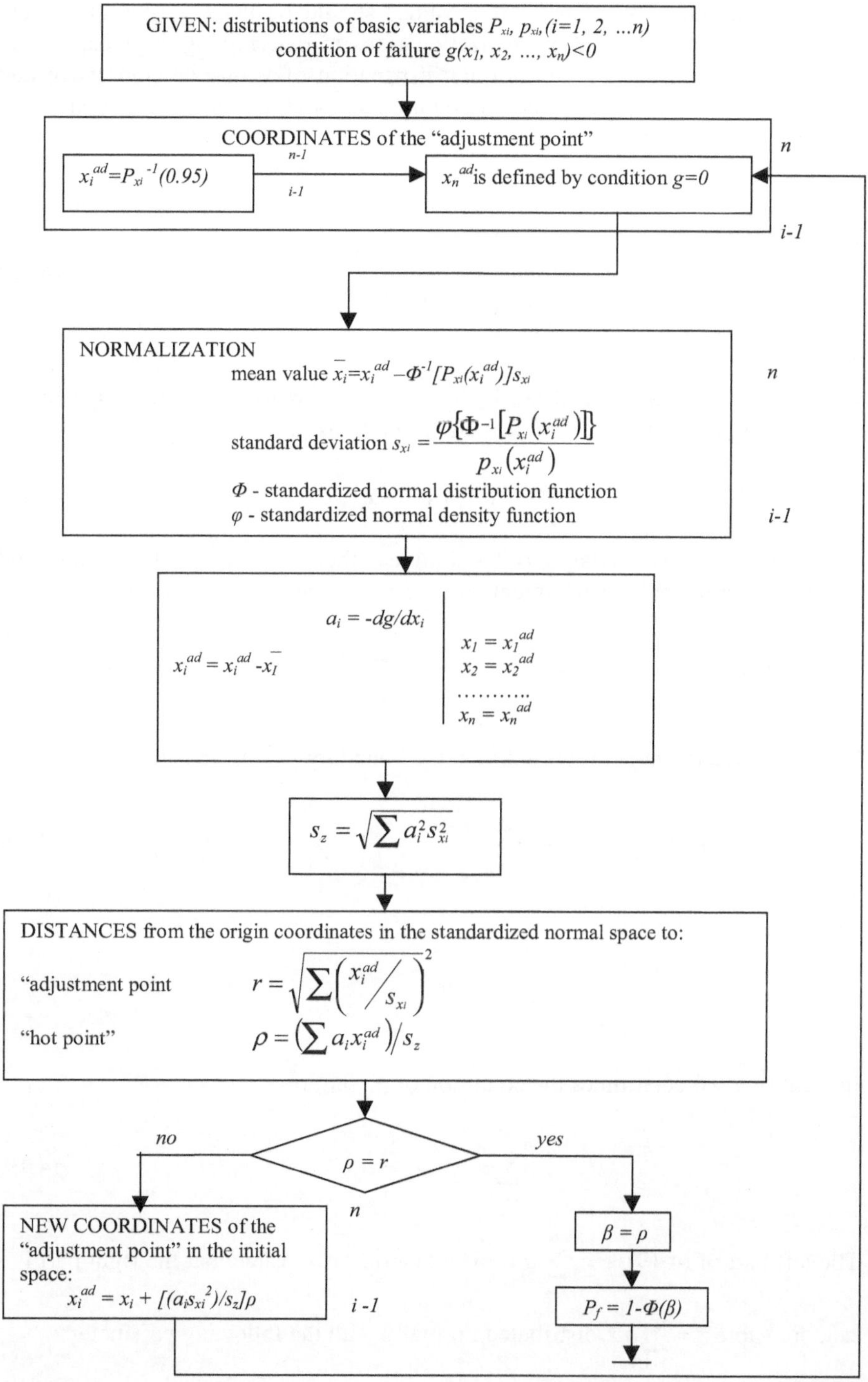

FIGURE 4.2 Diagram of "hot point" (FORM) method

Local normalization of the initial values should be made at the adjustment point. The process of local normalization leads to determining the mean $\overline{x}_1$ and standard deviation s_{xi} for such normal distribution of x_i that its cumulative and density functions at the adjustment point are equal to the corresponding values of the initial distribution x_i.

$$\left.\begin{aligned} &\overline{x}_i = x_i^{ad} - \Phi^{-1}[F_{xi}(x_i^{ad})s_{xi}] \\ &s_{si} = \varphi\left\{\Phi^{-1}\left[\frac{F(x_i^{ad})}{f_{xi}(x_i^{ad})}\right]\right\} \\ &i = 1 \div n \end{aligned}\right\} \tag{4.15}$$

where Φ and φ are integral and density functions of normal distribution, respectively.

Centering is conducted with the help of the transformation:

$$x_{ci}^{ad} = x_i^{ad} - \overline{x}_1 \tag{4.16}$$

If the failure region boundary $g = 0$ is nonlinear, then the limit state function should also be linearized at the adjustment point. This can be written as follows

$$h(x_1, x_2, ..., x_n) = g(x_1^{ad}, x_2^{ad}, ..., x_n^{ad}) + \sum_{i-1}^{n} \frac{\partial g}{\partial x_i} |(x_i = x_i^{ad}; ...; x_n = x_n^{ad}) \tag{4.17}$$

Since the adjustment point is chosen at the boundary $g = 0$, we get:

$$h(x_1, x_2, ..., x_n) = -\sum_{i=1}^{n} a_i(x_i - x_i^{ad}) = \sum_{i=1}^{n} a_i x_i^{ad} - \sum_{i-1}^{n} a_i x_i \tag{4.18}$$

where

$$a_i = -\frac{\partial g}{\partial x_i} |(x_i - x_i^{ad})$$

Inequality $h > 0$ determines the condition of no failure;

$$\sum_{i=1}^{n} a_i x_i^{ad} > \sum_{i=1}^{n} a_i x_i \tag{4.19}$$

The left part of (4.19), $c = \sum_{i-1}^{n} a_i x_i^{ad}$ is a deterministic value, but the right part is a random value $z = \sum_{i=1}^{n} a_i x_i$ distributed normally with the following parameters:

$$\text{mean value and standard deviation } \bar{z} = \sum_{i=1}^{n} a_i \bar{x}_i, s_z = \sqrt{\sum_{i=1}^{n} a_i^2 s_{x_i}^2} \tag{4.20}$$

Probability of no failure is expressed by formula

$$P_s = \int_{-\infty}^{\sum_{i=1}^{n} a_i x_i} f_z(z)dz = \Phi(\rho) \tag{4.21}$$

$$\text{where} \quad \rho = \frac{c - \bar{z}}{s_z} = \frac{\sum_{i=1}^{n} a_i (x_i^{ad} - \bar{x}_i)}{s_z} = \frac{\sum_{i=1}^{n} a_i x_{0i}^{ad}}{\sqrt{\sum_{i=1}^{n} a_i^2 s_{x_i}^2}} \tag{4.22}$$

Standardization of normalized initial values begins with making the transformation of $\bar{x}_1 / s_{xi}$, then in the space of these standardized normalized values the distance between the center of distribution (origin of coordinates) and linearized boundary of failure region (the "hot point") $h = 0$, will be equal to ρ. The hot point will be the crossing point of the perpendicular from the distribution center and straight line $h = 0$. Iterations will continue as long as the next adjustment point will coincide with the "hot point," i.e., distance r between

$$r = \sqrt{\sum_{i=1}^{n} \left(\frac{x_{0i}^{ad}}{s_{x_i}} \right)^2} \tag{4.23}$$

the adjustment point in the space of standardizing normal values and the distribution center will to be equal to ρ. This means that the linearization of a failure region boundary was realized at the point with maximum density of jointly distributed initial values at this boundary and the inaccuracy due to normalization and linearization is small, that is, reliability index $\beta = \rho$, and probability of no failure $P_s = \Phi(\beta)$. If a consecutive iteration results in $\rho \neq r$, then the adjustment point for the next iteration can be found by the following way: all its coordinates, expect one only, are taken equal to the appropriate hot point's coordinates from preceding iteration, and the last coordinate will be found from the condition $g = 0$. Thus:

$$x_i^{ad} = \bar{x}_i + \left(a_i s_{xi}^2 / s_z \right) \rho \tag{4.24}$$

Parameters of lognormal distribution for the yield stress can be taken from Equation (4.10), and $t = 0.1$, $m = 19.51$. Functions $F(x)$ and $f(x)$ can be taken from tables of normal distributions with the attendant transformations: $u = (\ln \sigma_y - 19.51) / 0.1$; $F\sigma_y(x) = \Phi[u(x)]$; $f_{\sigma_y}(x) = 10\varphi[u(x)] / x$, where Φ and

φ are, respectively, integral and density distribution functions of the standardized normal law. Limit state function and its derivatives are given in (4.11) and (4.12).

1st Iteration

Adjustment point coordinates are calculated in accordance with Equation (4.14)

$$x_1^{ad} = F_{x1}^{-1}(0.95)$$

$$\exp\{-\exp[-(x_1^{ad} - 5 \times 10^5)/6 \times 10^4]\} = 0.95$$

Hence

$$\ln(-\ln 0.95) = -(x_1^{ad} - 5 \times 10^5)/6 \times 10^4$$

$$x_1^{ad} = 6 \times 10^4 - [\ln(-\ln 0.95)] + 5 \times 10^5 = 678200$$

Substituting x_1^{ad} into equality g = 0, we obtain

$$x_2^{ad} = 678200\{[60 \times 10^5/(1.2 \times 10^5 - 0.6782 \times 10^6)] + 208.33\} = 2190 \times 10^5$$

Numerical parameters can be obtained from relations (5.17):

$$\bar{x}_1 = 678200 - \Phi^{-1}(0.95)s_{x1} = 678200 - 1.645 s_{x1}$$

$$s_{x1} = \varphi(1.645)/f_{x1}(x_1^{ad})$$

$$f_{x1}(x_1^{ad}) = (1/6 \times 10^4)\exp[-(6.78 \times 10^5 - 5 \times 10^5)/6 \times 10^4]$$
$$\exp\{-\exp[-(6.78 \times 10^5 - 5 \times 10^5)/6 \times 10^4]\} = 8.123 \times 10^{-4}$$

$$\bar{x}_1 = 4.69 \times 10^5$$

$$s_{x1} = 1.27 \times 10^5$$

$\bar{x}_2$ $\bar{x}_2$ and s_{x2} are calculated analogously:

$$\bar{x}_2 = x_2^{ad} - \Phi^{-1}[(\ln x_2^{ad} - 19.51)/0,1]s_{x2}$$

$$s_{x2} = \varphi[(\ln x_2^{ad} - 19.51)/0.1] / (10/x_2^{ad})\varphi[(\ln x_2^{ad} - 19.51)/0.1] = x_2^{ad}/10 = 219\text{x}10^5$$

$$\bar{x}_2 = 0868\text{x}10^5$$

Centering procedures yield

$$x_{c1}^{ad} = 678200 - 469390 = 2.08 \times 105$$

$$x_{c2}^{ad} = 2190 \times 10^5 - 2868\text{x}10^5 = 678 \times 10^5.$$

In accordance with equation (4.20)

$$a_i = \partial g/\partial x_i,\ a_1 = 472.4,\ a_2 = -1$$

From (3.20)

$$s_z = \sqrt{\left(472,4\cdot 1,27\cdot 10^5\right)^2 + (219\cdot 10^5)^2} = 638,4\cdot 10^5$$

$$\sum_{i=1}^{n} a_i x_{ci}^{ad} = 472.4\times 2.08\times 10^5 + 678\times 10^5 = 1664.4\times 10^5$$

From Equation (4.22) we derive:

$$\rho = 1664.4/638.4 = 2.61$$

Using Equation (4.23) we obtain

$$r = \sqrt{\left(\frac{2,08}{1,27}\right)^2 + \left(\frac{678}{219}\right)^2} = 3,506 = 3.506$$

It is evident that r ≠ ρ, and the next iteration is necessary.

2nd Iteration
In accordance with (3.24) the adjustment point's coordinates are

$$x_1^{ad} = 4.69 \times 10^5 + [472.4 \times (1.27 \times 10^{52} \times 2.61]/638.4 \times 10^6 = 7.8 \times 10;$$

$$x_2^{ad} = 7.8 \times 10^5 \times [60/(1.2 - 0.78) + 208.33] = 2741.2 \times 10^5$$

Normalization is conducted next:

$$\bar{x}_1 = 7,8\cdot 10^5 - \Phi^{-1}\left[F_{x_1}\left(7,8\cdot 10^5\right)\right]s_{x_1}$$

$$F_{x_1}\left(7,8\cdot 10^5\right) = \exp\left[-\exp\left(-\frac{7,8\cdot 10^5 - 5\cdot 10^5}{6\cdot 10^4}\right)\right] = 0,99067$$

$$\Phi^{-1}\left(0,99067\right) = 2,346$$

$$f_{x_1}\left(7,8\cdot 10^5\right) = \frac{1}{6\cdot 10^4}\exp\left(-\frac{7,8\cdot 10^5 - 5\cdot 10^5}{6\cdot 10^4}\right)$$
$$\exp\left[-\exp\left(-\frac{7,8\cdot 10^5 - 5\cdot 10^5}{6\cdot 10^4}\right)\right] = 1,546\cdot 10^{-7}$$

$$s_{x_1} = \frac{\varphi(2,346)}{f_{x_1}\left(7,8\cdot 10^5\right)} = 1,646\cdot 10^5 \quad \bar{x}_1 = 7,8\cdot 10^5 - 2,35\cdot 1,65\cdot 10^5 = 3,94\cdot 10^5$$

$$\bar{x}_2 = 2741{,}2\cdot 10^5 - \Phi^{-1}\left[F_{x_2}\left(2741{,}2\cdot 10^5\right)\right]s_{x_2}$$

$$\Phi^{-1}\left[F_{x_2}\left(2741{,}2\cdot 10^5\right)\right] = 10\left(\ln 2741{,}2\cdot 10^5 - 19{,}51\right) = -0{,}84$$

$$s_{x_2'} = 274{,}1\cdot 10^5; \bar{x}_2 = \left(2741{,}2 + 0{,}84\cdot 274{,}1\right)\cdot 10^5 = 2971{,}2\cdot 10^5$$

Centering procedure yields

$$x_{01}^{ad} = 7{,}8\cdot 10^5 - 3{,}93\cdot 10^5 = 3{,}86\cdot 10^5$$

$$x_{02}^{ad} = 2741{,}2\cdot 10^5 - 2971{,}2\cdot 10^5 = -230\cdot 10^5$$

Parameters equal

$$a_1 = 616.63\ a_2 = -1$$

The final step consists of determination of r and ρ.

$$s_z = \sqrt{[(616.63 \times 1.65 \times 10^5)^2 + (274 \times 10^5)^2]} = 1051.67 \times 10^5$$

$$\sum_{i=1}^{2} a_i x_{ci}^{ad}\ {}_i x_{ci}^{\ ad} = 616.63 \times 3.86 \times 10^5 + 230 \times 10^5 = 2611.9 \times 10^5$$

$$\rho = 2611.9 \times 105/1051.67 \times 105 = 2.48$$

$$r = \sqrt{[(3.86/1.65)2 + (230/274.1)2]} = 2.49$$

One can see that $r \approx \rho$ and as $\rho = \beta$, then $P_f = 0.0064$ and reliability is $P_s = 0.9936$.

This iteration procedure is illustrated in Figure 4.3.

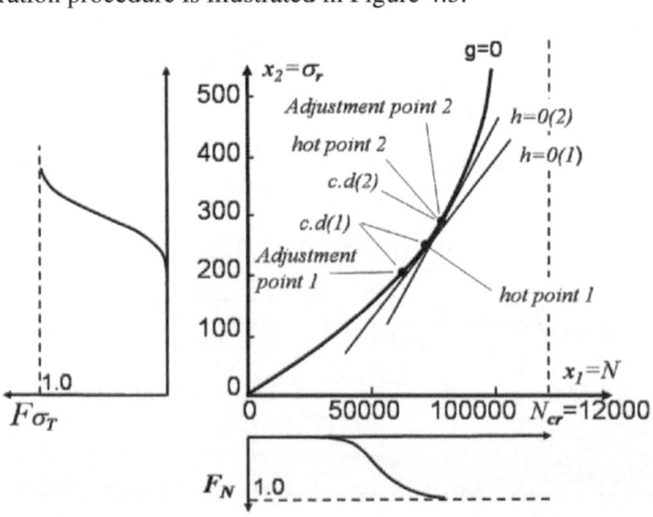

FIGURE 4.3 σ_y–N relationship

4.3 MONTE-CARLO METHOD

4.3a Monte-Carlo Technique

Within the context of this method the integral $\int_0^\infty P_g(\xi)p_F(\xi)d\xi$ is used as the definition of function P_R, and its estimates are derived by the random sample average, that is:

$$\bar{P}_f = \int_0^\infty P_R(x)dx = \bar{P}_R(F) \approx \frac{1}{m}\sum_{i-1}^{m} P_R(F_i) \tag{4.25}$$

where m is the number of trials.

The calculation procedure is as follows. At each trial, realization F_i of variable F is generated in accordance with its probability density function (this function is supposed to be known) and the value of the distribution function of quantity R, corresponding to argument F_i, is determined (this distribution function is supposed to be known as well). Then the average of these values from all trials is calculated. If derivation of the density function of quantity F is a difficult task due to its dependence on many basic variables (x_1, x_2, …, x_n), then at each trial the realizations x_{ji} of x_i are generated in accordance with their density functions at the time when values F_i and $P_R(F_i)$ are calculated. Let's illustrate obtaining a numerical solution by Monte-Carlo procedure using the same cross section as in Figure 4.1.

There is

$$P_f \cong \frac{1}{m}\sum_{i=1}^{m} P_{\sigma_T}(N_i).$$

The generator of random numbers should simulate the value of axial force N with uniform distribution in the interval [0, 1], with the accompanying transformation of distribution function $F_N(n)$ as shown in Figure 4.4.

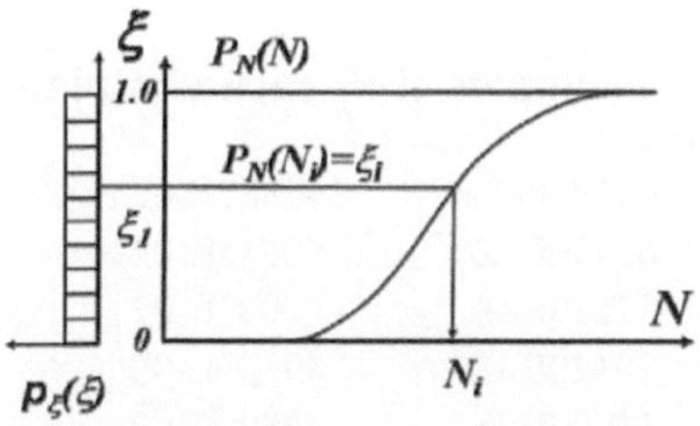

FIGURE 4.4 Probability distribution function of axial force N

The generator releases the first random number $\xi = 0.084$, that is: $F_N(N_1) = \exp[-\exp(-5 \times 10^5/6 \times 10^4)] = 0.084$; $N_1 = 6 \times 10^4[-\ln(-\ln 0.084)] + 5 \times 10^5 = 4.4549 \times 10^5 N$.

The yield stress can be found from Equation (4.5):

$$\sigma_{y1} = N_1[(60 \times 10^6/1.210^6 + 208.3] = 1281.71 \times 10^5\ Pa$$

The value of the distribution function coincides with the value of yield stress. $F_{\sigma_y}[\sigma_{y1}(N_i)] = \Phi[10(\ln \sigma_{y1} - 19.51)]$

This procedure, performed 1000 times, yields the mean value $\bar{P}_f = \frac{1}{1000}\sum_{i=1}^{1000} F_{\sigma_y}(N_i) = 0.0049$

This result tells nothing about its accuracy and trustworthiness. In other words, the question arises whether 1000 tests are enough or not. Necessary procedures in all simulations methods consist in estimating their trustworthiness; 20 samplings with 1000 tests in each are made and 20 values of the failure probability $\bar{P}_f$ are obtained. They are considered as random variable with mean value equal to $\bar{P}_f$. As the mean-root-square deviations of this random value are known, then for definition of a confidence interval P_f TEST (Student's criterion) was used

$$v = \pm t_p \sqrt{\frac{\sum_{i=1}^{M} P_{f_i}^2 - M\bar{P}_{f_i}^2}{M(M-1)}}$$

Different values of t_p are associated with different values of confidence probability $P(t)$. In the given case $M = 20$ the desired probability P_f is estimated as

$$P_f = \bar{P}_{fi} \pm t_p \sqrt{\frac{\sum_{i=1}^{20} P_{f_i}^2 - 20\bar{P}2_{f_i}}{20 \cdot 19}}$$

The estimates of probabilities P_f of yield stress occurrence in the column's outermost fiber are presented in Table 4.1. Values of confidence intervals for different

TABLE 4.1
Twenty sample estimates of P_{fi} each containing 1000 trials

85333E-02	64393E-02	50887E-02	46491E-02
67381E-02	63440E-02	50115E-02	42542E-02
66616E-02	57279E-02	50007E-02	37344E-02
66438E-02	55120E-02	49169E-02	33796E-02
66137E-02	53909E-02	48898E-02	10305E-02

TABLE 4.2
Confidence intervals for the given confidence probabilities

Confidence probability	Confidence interval	Confidence probability	Confidence interval
0.99000	35525E-03	0.93000	33372E-03
0.98000	35167E-03	0.92000	33013E-03
0.97000	34808E-03	0.91000	32655E-03
0.96000	34449E-03	0.90000	32296E-03
0.95000	34090E-03	89000	31737E-03
0.94000	33731E-03	0.88000	31578E-03

values of confidence probability are given in Table 4.2. Each confidence interval estimate was derived using the TEST criteria.

The mean value $\overline{P}_{fi}$ estimated with respect to 20000 tests is equal to 0.0058. For confidence probability $P(t) = 0.95$, P_f varies within the interval of $P_f = 0.00579 \pm 0.00034$, that is, $0.00613 > P_f > 0.00545$. If the confidence interval range is too wide, the sample volume or their number should be increased. Mean value of samples is $P_{fi} = 57317\text{E-}02$.

4.3b Monte-Carlo Method for Stratified Modeling Samples

There are some modifications of the Monte-Carlo method which increase its efficiency due to a decrease in the estimation's depression. On the ways for decreasing the estimation's dispersion in stratifying the modeling samples, for example, samples of F_i when using Equation (4.25). The integration interval is divided into n sub-intervals:

$$P_f = \int_0^{\infty} P_R(x)p_F(x)dx = \sum_{j=1}^{n} \int_{F_{j-1}^*}^{F_i^*} P_R(x)p_F(x) = \sum_{j=1}^{n}\left(\int_{F_{j-1}^*}^{F_i^*} p_F(x)dx \right)$$
$$\left[\overline{P}_R\left(F_j\right)\right]_j = \sum_{j=1}^{n} p_{F_j}\left[\overline{P}_R\left(F_j\right)\right] \approx \sum_{j=1}^{n} p_{F_j} \frac{1}{m} \sum_{i=1}^{m_j} P_R\left(F_{ij}\right) \tag{4.26}$$

Where p_{Fi} is the probability of the value of F belonging to the jth sub-interval; $\overline{P_R}\left(F_j\right)$ is the mean value of the function $P_R(F)$ provided F belongs to the jth sub-interval $F_{j-1}^* < F_j < F_j^*$; m_j is the number of realizations within the jth sub-interval; F_{ij} is the realization of random variable F within the jth sub-interval.

It is obvious that the deviation of estimates essentially decreases in this case, as it is a sum of small deviations within a single interval. However, when small probabilities are calculated, the sub-intervals, making the main contribution in estimate (4.26), are in the region with small probable F values. Thus, a considerable part of samples becomes unneeded, and the required number of realizations will be rather high.

The modified Monte-Carlo method cures this defect. Instead of two steps (first, making a modeling of sample for F values, and second, making its stratification), in the modified method, a stratified sample base is formed at once at necessary sub-intervals with the given volume of sub-samples. Limited random values are modeled for the realization of this idea, $P_{j-1}^{*}<P_j<P_j^{*}$, with density distribution function

$$P_F(x)=\frac{p_{Fj}(x)}{P_{Fj}} \tag{4.27}$$

Let's continue the numerical example of the bar in Figure 4.1.

We have

$$P_f\approx\sum_{j=1}^{n}p_{N_j}\frac{1}{m}\sum_{i=1}^{m}P_{\sigma y}\left[\sigma_y\left(N_{ij}\right)\right]$$

First, the integration interval (N_{left}, N_{right}) should be found. The right boundary can be determined from the following conditions

$$P_N(N_{right})=0.9999$$

$$N_{right}=P_N^{-1}(0.9999)=6\times10^4[-\ln(-\ln 0.9999)+5\times10^5]=10.5\times10^5$$

The conditions for determining the left boundary are: $P_{\sigma y}[\sigma_y(N_{left})]=0.0001$

$$P_{\sigma y}[\sigma_y(N_{left})]=\Phi\{10[ln\sigma_y(N_{left})-19.51]\}=-3.72;\ \sigma_y(N_{left})=2055$$

From (3.11) with $g=0$, we get:

$$N_l\left(\frac{60\cdot10^6}{1{,}2\cdot10^6-N_l}+0{,}0208\right)=2055\cdot10^5.$$

$$\mathrm{N}_{\mathrm{left}}=6.5\times10^5\ \mathrm{H}$$

The integration interval will be divided into four sub-intervals $(j=1,2,3,4)$ with the left boundaries equal to, respectively, $N_1=N_{left}=6.5\times10^5$, $N_2=7.5\times10^5$, $N_3=8.5\times10^5$, $N_4=9.5\times10^5$.

Distribution functions are determined at the interval boundaries

$$P_N(N_1)=\exp\left(-\exp\frac{6.5\times10^5-5\times10}{0.6\times10^5}\right)^5=0.9212$$

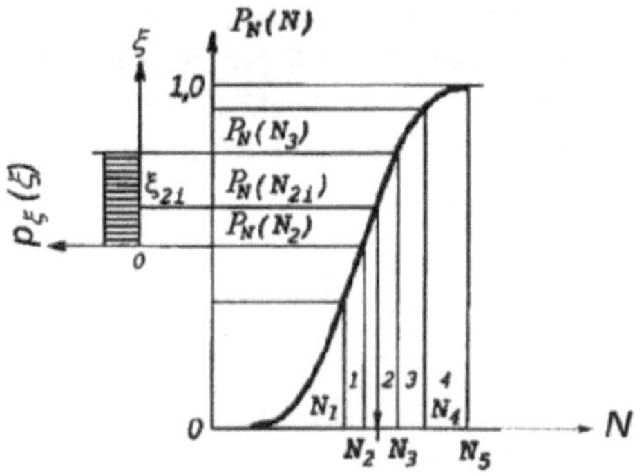

FIGURE 4.5 Probability distribution function of axial force N and sub-intervals

Similarly,

$$P_N(N_2) = 0.9846,\ P_N(N_3) = 0.9971,\ P_N(N_4) = 0.9994,\ P_N(N_5) = 0.9999 = N_{right}$$

Occurrence probabilities that load values are within the sub-intervals are equal to:

$$P_{N1} = P_N(N_2) - P_N(N_1) = 0.0634,\ P_{N2} = 0.0125,\ P_{N3} = 0.0023,\ P_{N4} = 0.0005$$

Samples with m entries are modeled for each sub-interval due to random number generator (Figure 4.5). The random value of N_{ij} for the jth sub-interval can be determined as a function of the random number ξ_{ij}

$$N_{ij} = P_N^{-1}\{P_N(N_j) + [P_N(N_{j+1}) - P_N(N_j)]\xi_{ij}\} = P_N^{-1}[P_N(N_j) + P_{Nj}\xi_{ij}] = P_N^{-1}[P_N(N_{ij})]$$

Yield stress $\sigma_y(N_{ij})$ is determined from the condition $g = 0$ in equation (3.5):

$$\sigma_y\left(N_{ij}\right) = N_{ij}\left(\frac{60\times10^5}{1.2\times10^6 - N_{ij}} + 208.3\right)$$

and distribution function is:

$$P_{\sigma y}[\sigma_y(N_{ij})] = \Phi\{10\ln\sigma_y(N_{ij}) - 19.51]\}.$$

This procedure is repeated m times for each sub-interval. The failure probability P_f is estimated using the formula at the beginning of this example. Accuracy at the given certainty can be analyzed as in the regular Monte-Carlo method, see the previous example.

4.4 SIMULATION METHOD

If the probability is estimated by the frequency of the event ($F>R$), then a sufficiently large number of statistical tests are performed according to the Bernoulli scheme, that is, random realizations of all the initial values are generated at each test. The values of the quantities F and R are determined as functions of these realizations and the condition $F>R$ is checked. If the condition is met, the test result is considered a failure. The frequency ν of failure occurrence is considered as an estimate of its probability P_f:

$$\nu = \frac{k}{m} \approx P_f \tag{4.28}$$

where k is the number of failures, m is the total number of tests.

The method is extremely simple and universal, but it requires a mandatory analysis of the proximity of the estimate ν to the desired probability P_f, which depends on the number of tests m. Known methods of such analysis are based on limit theorems:

- Bernoulli, which states that for $m \to \infty$, $\nu \to P_f$ and that there is an asymptotically normal distribution.
- Khinchin (the law of large numbers), which states that the average frequencies ν at tend $m \to \infty$ to its mathematical expectation.
- Landsberg-Levy (central limit theorem), which states that the average frequency ν has an asymptotically normal distribution.

However, the addition theorems state that the binomial distribution (and the number of failures is distributed exactly according to the binomial law) is stable when adding independent variables (self-reproduces) and that the sum of mutually independent random variables has a normal distribution if and only if they are all normally distributed. Therefore, the question arises: at what total number of tests can we use limit theorems and accept asymptotic distributions to construct practical procedures for checking the accuracy and assurance of the obtained estimates. It is almost impossible to answer this question. In addition, the known procedures use an asymptotic frequency ν distribution and are intended to build a confidence value for its mathematical expectation interval, rather than the desired probability. These procedures are quite complex and not sufficiently convincing, which often leads to the rejection of the analysis for proximity of ratings. In the book by Raizer (2010), fairly simple procedures for constructing a confidence interval for the probability of failure are presented, using the distribution of this probability. The essence of the procedure is as follows.

The number of k failures in m tests with the probability of failure $p = P_f$ in one test is distributed according to the binomial law

$$f_{k/p,m}(k) = C_m^k p^k (1-p)^{m-k} \tag{4.29}$$

This distribution is discrete, since k is an integer. In the real situation, we know k and m and we don't know p. We can only assume with some probability what the p value was in each of the tests performed. This assumption is characterized by the density of the probability distribution of the value p, which is expressed from (3.29), if it takes as an argument not k, but p and introduces a normalizing factor χ:

$$f_{p/k,m}(p) = \chi C_m^k p^k (1-p)^{m-k} \tag{4.30}$$

The normalizing factor is determined from the condition $\int_{-\infty}^{\infty} f_p(p)dp = 1$ and is equal to $(m+1)$. With this in mind, the formula for the probability density function (4.29) is presented as

$$f_{p/k,m}(p) = (m+1)C_m^k p^k (1-p)^{m-k} \tag{4.31}$$

And represents the beta distribution (continuous distribution)

$$f_p(p) = \frac{(\alpha+\beta-1)!}{(\alpha-1)!(\beta-1)!} p^{\alpha-1}(1-p)^{\beta-1} \tag{4.32}$$

for $(\alpha-1) = k; (\beta-1) = m-k$

To obtain the bounds of the confidence interval for p, it is necessary to calculate the values of the distribution function F_p for the beta distribution, which is an extremely difficult task. Tables of this function are available only for small values of m and k. However, by elementary but cumbersome transformations, it can be proved that

$$F_{p/k,m}(p) = 1 - F_{k/h,(m+1)}(m) \tag{4.33}$$

where $F_{p/k,m}$ – is the distribution function p for the given m and k (beta distribution);

$F_{k/p,m}$ – probability distribution function k for given m and p (binomial distribution).

To calculate the binomial distribution function for large m, small p (p<0,1) and finite m, p, one can use the Poisson distribution.

$$f_{r/p,m}(k0 = C_m^k p^k (1-p)^{m-k} \approx \frac{(mp)^k}{k!} e^{-mp} \tag{4.34}$$

In the considered case

$$1 - F_{p/k,m}(p) = F_{k/p,(m+1)}(k) = \sum_{i=0}^{k} e^{-\lambda} \frac{\lambda^i}{i!} = \sum_{i=0}^{k} f_k(i) \tag{4.35}$$

where $\lambda = (m+1)$, and $f_k(i)$ is calculated by the recurrent formula

$$f_k(i) = \frac{\lambda}{i} f_k^{(i-1)}; f_k(0) = e^{-\lambda} \tag{4.36}$$

In practice, a function $F_{\lambda/k}(\lambda)$ is constructed from which with the transformation $\lambda = (m+1)k$ one can get a function $F_{p/k}(p)$ for any m. This function can be used to determine the inverse of a probability distribution $\gamma : \lambda_\gamma = F_{\lambda/k}^{-1}(\gamma)$ and confidence coefficients

$$\eta_\gamma(k) = \frac{\lambda_\gamma}{k}$$

The product $\eta_\gamma \nu_1$ is the boundary of the confidence interval corresponding to the confidence γ, where

$$\nu_1 = \frac{k}{m+1}.$$

For large m and small k, you can take $\nu_1 \approx \nu$. Table 4.3 (Raizer, 2010) shows the values of $\eta_{0,95}$ and $\eta_{0,99}$, depending on the number of failures k. The practical application of this method can be demonstrated in the following example.

Let there be 500 tests with k = 20 cases of failures. Frequency of failures in that case Then, with a confidence probability of 0.95, we can say that $P_f \le \eta_{0,95}\nu = 1{,}453 \cdot 0{,}004 = 0{,}005812$ and with a probability of 0.99 $P_f \le \eta_{0,99}\nu = 1{,}655 \cdot 0{,}004 = 0{,}00622$.

The confidence coefficients η depend only on the number of recorded failures k and do not depend on the number of tests. This makes it possible to make a complete table of the practically necessary values of the coefficients. The procedure for analyzing the results of statistical modeling becomes extremely simple and is expressed by the formula,

$$P_{rob}\{P_f \le \eta_\gamma \nu_1\} = \gamma \tag{4.37}$$

You can set the desired confidence factor (i.e., accuracy and confidence) and run the tests until the appropriate number of failures is obtained. This solves the problem of determining the required number of tests. As you know, theoretically, the true value of any characteristic that determines a random variable can be considered for an infinitely large N-number of tests. In engineering studies of building structures, where high levels of reliability of the order of 0.999–0.9999 are required, one has to deal with very small values of the probability of failure and investigate the "tails" of the distribution function. These "tails" are determined with sufficient accuracy for N at least $10^6 \div 10^9$ (depending on the type of problem).

The main difficulty of the problem solved by the method of statistical tests is the large volume of calculations performed. The general algorithm for structural analysis by statistical testing method is as follows. All parameters of the load-bearing

TABLE 4.3
Confidence coefficients

κ	$\eta_{0,95}$	$\eta_{0,99}$	κ	$\eta_{0,95}$	$\eta_{0,99}$
1	4,74390	6,63900	31	1,34961	1,50353
2	3,14790	4,20300	32	1,34320	1,49414
3	2,58460	3,34850	33	1,33714	1,48532
4	2,28840	2,90137	34	1,33134	1,47688
5	2,10261	2,62200	35	1,32583	1,46884
6	1,97375	2,42850	36	1,32058	1,46117
7	1,87832	2,28579	37	1,31556	1,45382
8	1,80435	2,17537	38	1,31076	1,44683
9	1,74503	2,08717	39	1,30615	1,44012
10	1,69623	2,01450	40	1,30174	1,43370
11	1,65524	1,95368	41	1,29750	1,42756
12	1,62025	1,90175	42	1,29343	1,42161
13	1,58990	1,85688	43	1,28951	1,41593
14	1,56332	1,81768	44	1,28574	1,41044
15	1,53981	1,78290	45	1,28211	1,40513
16	1,51884	1,75191	46	1,27862	1,40005
17	1,49996	1,72412	47	1,27522	1,39513
18	1,48288	1,69900	48	1,27197	1,39038
19	1,46733	1,67613	49	1,26880	1,38582
20	1,45312	1,65525	50	1,26574	1,38135
21	1,44002	1,63600	51	1,26279	1,37706
22	1,42795	1,61823	52	1,25992	1,37290
23	1,41676	1,60180	53	1,25714	1,36888
24	1,40636	1,58656	54	1,25185	1,36500
25	1,39664	1,57236	55	1,25185	1,36118
26	1,38756	1,55902	56	1,24929	1,35750
27	1,37906	1,54656	57	1,24683	1,35389
28	1,37105	1,53488	58	1,24443	1,35045
29	1,36350	1,52379	59	1,24210	1,34705
30	1,35635	1,51340	60	1,23983	1,33879

capacity R_1, ... ,R_n and actions F_1,...,F_n that have variability are considered as the random variables with known laws of distribution. The distribution laws are set numerically using random number generators. A deterministic calculation method is also assumed to be known. Failure of the structure is considered to be non-compliance with the accepted basic design condition. The determination of the failure rate and the construction of a histogram of the strength reserve function is carried out according to the following scheme.

- By the method of statistical modeling, according to the known laws of distribution, n realizations of random variables are assigned - ($I = 1, \ldots$, n). The number of tests n is assigned based on the required time.
- Deterministic design calculations n are performed according to the selected method. As a result of calculations, the value of the strength reserve function g defined as $R - F$ (Raizer, 1986) is determined n times.
- If the $g \leq 0$ value is fixed, a failure is detected. The failure rate is calculated using the formula (4.28).
- The obtained frequency is compared with the predefined failure probability P_f. If one of the parameters changes, for example, the percentage of reinforcement (when designing reinforced concrete structures), and you should return to the second step. Iteratively, the calculations are repeated until the desired failure rate is reached.
- When the specified failure rate is reached, histograms of the relative frequencies of the strength reserve distribution function are plotted for the obtained values over an interval equal to m standards *of* g. In this case, the right and left boundaries of the interval lie at a distance of *m/2* standards to the right and left of the mathematical expectation $\bar{g}$.

The method of statistical tests with sufficient accuracy allows you to determine the estimate of the probability of failure of structures. The approach used to determine the upper limit of the confidence interval for the value of the failure probability estimate is quite simple and effective. The method of constructing the confidence interval described here allows you to normalize not the required number of tests, but the required number of failures. Then, according to this number of failures, in agreement with Table 4.3 you can get the coefficient $\eta_{0,95}$ or $\eta_{0,99}$, by which you need to multiply the obtained probability of failure to get the upper limit of the confidence interval, which will not be exceeded with a probability of 0.95 and 0.99, respectively.

4.5 DIRECT INTEGRATION OF DISTRIBUTION FUNCTION

Two random values X_1 and X_2 with marginal distribution functions $F_1(x_1)$, $F_2(x_2)$ and density functions $f_1(x_1)$, $f_2(x_2)$ are considered in this section. Failure region's boundary shown at Figure 4.6 is given by:

$$X_2 = \varphi(X_1) \tag{4.38}$$

Then, the reliability is defined as the probability that the random point with coordinates (X_1, X_2) will belong to region 4. If $f(x_1, x_2)$ is a joint density function of random variables X_1 and X_2, then for the joint distribution function $F(x_1, x_2)$ can be defined by the following equation:

$$F(x_1, x_2) = \int_{-\infty}^{x_1} \int_{-\infty}^{x_2} f(x_1, x_2) dx_1 dx_2 \tag{4.39}$$

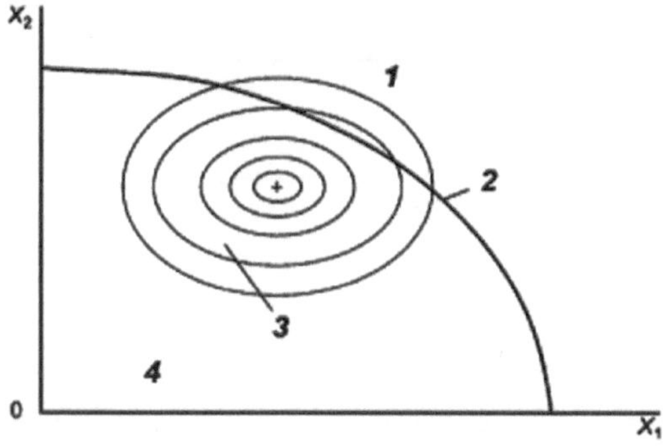

FIGURE 4.6 Boundary of failure region; 1. Failure region; 2. Boundary of failure region; 3. Compatible density function; 4. Safe region

For a system of independent random variables

$$f(x_1, x_2) = f_1(x_1) f_2(x_2) \tag{4.40}$$

and Equation (4.39) becomes:

$$F(x_1, x_2) = \int_{-\infty}^{x_1} \int_{-\infty}^{x_2} f_1(x_1) f_2(x_2) dx_1 dx_2 \tag{4.41}$$

Alternately, using distribution functions of random variables, we get

$$F(x_1, x_2) = F_1(x_1) F_2(x_2) \tag{4.42}$$

It should be noted that the joint distribution function for the system of two random variables (X_1, X_2) is the probability of simultaneous realization of two inequalities $X_1 < x_1$ and $X_2 < x_2$:

$$F(x_1, x_2) = P[(X_1 < x_1) \cap (X_2 < x_2)] \tag{4.43}$$

In a geometric interpretation, distribution function $F(x_1, x_2)$ is the probability of a random point with coordinates (X_1, X_2) landing in an infinite quadrant with its vertex at (x_1, x_2), see Figure 4.7. Probability of a random point (X_1, X_2) landing in rectangular region R, bounded with abscissas α and β and ordinates γ and δ, can be expressed via the joint probability distribution function of the considered system (Figure 4.8).

Indeed, the probability of landing in the rectangular region R is equal to the probability of landing to the quadrant with vertex (β, δ), minus the probability of landing to the quadrant with vertex (α, δ), minus the probability of landing to

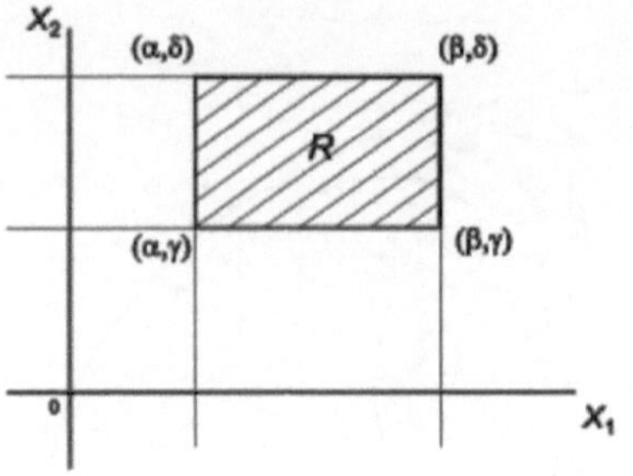

FIGURE 4.7 Geometric interpretation of the joint distribution function

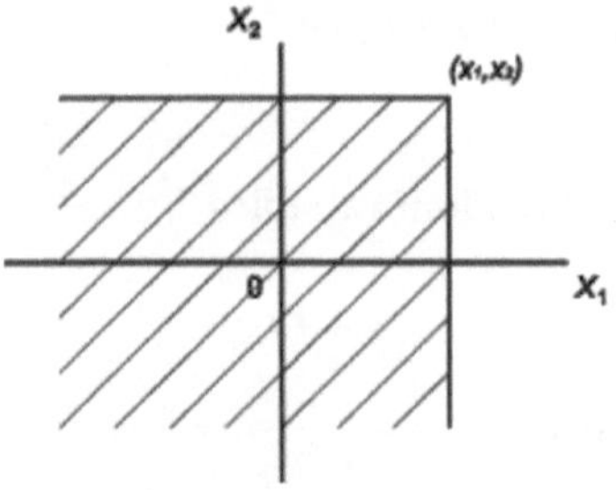

FIGURE 4.8 Random point in rectangular region

the quadrant with vertex (β, γ), plus the probability of landing into quadrant with vertex (α, γ). As a result, the following formula is derived:

$$P\big((X_1,X_2)\subset R\big)=F(\beta,\delta)-F(\alpha,\delta)-F(\beta,\gamma)+F(\alpha,\gamma) \tag{4.44}$$

In view of (4.42), Equation (3.44) is transformed into:

$$P\big((X_1,X_2)\subset R\big)=F(\beta)F(\delta)-F(\alpha)F(\delta)-F(\beta)F(\gamma)+F(\alpha)F(\gamma) \tag{4.45}$$

The safe region is divided into n rectangles with equal steps h along the X_1 axis (Figure 4.9).

Reliability, which equals to the probability of random point landing in the region of no failure D, can be found as a sum of probabilities of landing in each of the rectangular regions D_n:

$$P\big((X_1,X_2)\subset D\big)=P\big((X_1,X_2)\subset D_1\big)+P\big((X_1,X_2)\subset D_2\big)+\ldots+P\big((X_1,X_2)\subset D_n\big) \tag{4.46}$$

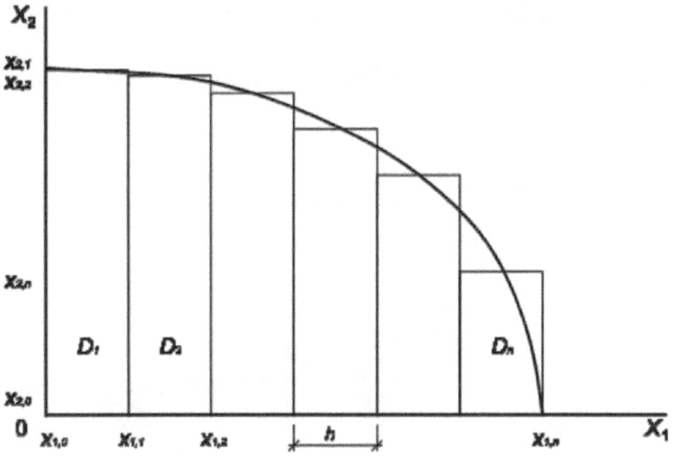

FIGURE 4.9 The probability of random point landing in arbitrary region

Using equation (4.44), (4.46) is transformed into:

$$\begin{aligned} P\big((X_1,X_2)\subset D\big) = F(x_{1,1},x_{2,1}) - F(0,x_{2,1}) - F(x_{1,1},0) + F(0,0) + F(x_{1,2},x_{2,2}) - \\ -F(x_{1,1},x_{2,2}) - F(x_{1,2},0) + F(x_{1,1},0) + \ldots + \\ +F(x_{1,n},x_{2,n}) - F(x_{1,n-1},x_{2,n}) - F(x_{1,n},0) + F(x_{1,n-1},0). \end{aligned} \tag{4.47}$$

If the considered random variables take only positive values (strength or applied loads, e.g.), then the probability of random point landing in a region with negative values is zero and Equation (4.47) is transformed into:

$$\begin{aligned} P\big((X_1,X_2)\subset D\big) = F(x_{1,1},x_{2,1}) - F(0,x_{2,1}) + F(x_{1,2},x_{2,2}) - F(x_{1,1},x_{2,2}) + \ldots + \\ +F(x_{1,n},x_{2,n}) - F(x_{1,n-1},x_{2,n}) \end{aligned} \tag{4.48}$$

or

$$P\big((X_1,X_2)\subset D\big) = \sum_{i=0}^{n}\big(F(x_{1,n},x_{2,n}) - F(x_{1,n-1},x_{2,n})\big) \tag{4.49}$$

Taking into account (3.41), we get:

$$P\big((X_1,X_2)\subset D\big) = \sum_{i=0}^{n}\Big[\big(F(x_{1,i}) - F(x_{1,i-1})\big)\cdot F(x_{2,i})\Big] \tag{4.50}$$

The more is the "*i*" value, the higher accuracy is in computation of reliability P_s. The exact value of P_s is obtained when *i* tends to infinity. Lower and upper estimates of P_s can be derived if the region D is divided into rectangles as shown at Figure 4.10.

The reliability can be similarly calculated in the case of three random variables X_1, X_2, X_3. Safe region is a volume bounded with a surface of no failure and coordinate planes $X_1 0 X_2$, $X_1 0 X_3$, and $X_2 0 X_3$. Region D is divided into rectangular parallelepipeds (Figure 4.11).

It is obtained by n partitions along the X_1 axis and m partitions along the X_2 axis. The probability of a random point landing into D region equals to the sum of

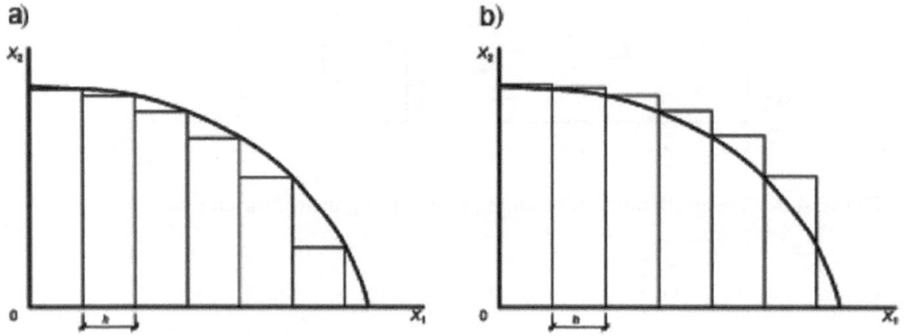

FIGURE 4.10 Evaluation of the probability of a random point landing in an arbitrary region: a) from below, b) from above

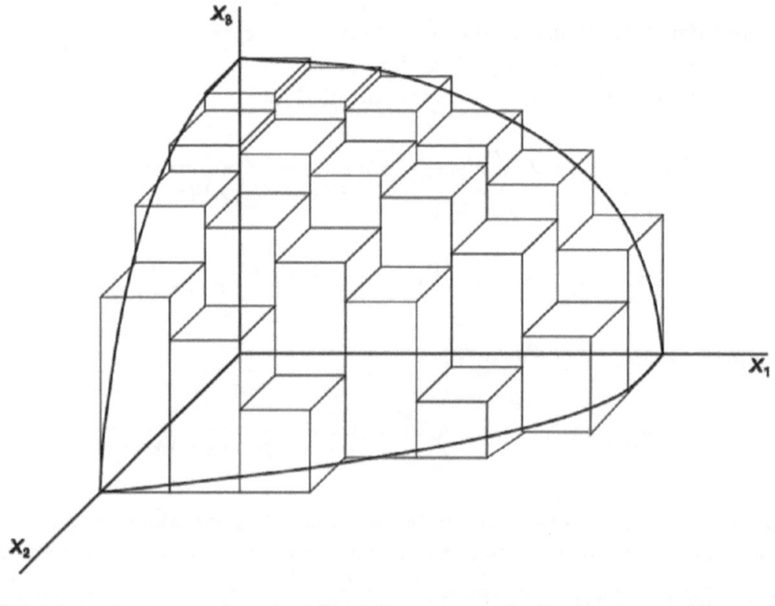

FIGURE 4.11 Evaluating the probability of a random point landing in an arbitrary volume

the probabilities of a point landing in each parallelepiped. After manipulations we obtain:

$$P\left((X_1,X_2,X_3)\subset D\right)=\sum_{i=0}^{n}\sum_{j=0}^{m}\left((F(x_{1,n})-F(x_{1,n-1}))\cdot(F(x_{2,m})-F(x_{2,m-1}))\cdot F(x_{3,ij})\right) \tag{4.51}$$

where $x_{3,ij}$ is altitude of corresponding parallelepiped (coordinate on the X_3 axis).

4.6 INFLUENCE OF FAILURE BOUNDARY CURVATURE ON RELIABILITY

As mentioned before, the hot point, FOSM, and SORM methods are widely used in reliability problems. A drawback of these methods is in the linearization procedure of the limit state function at "hot point" of failure region boundary. Inaccuracy with regard to this linearization should be evaluated. Following Veneziano (1978), three possible cases for the failure surface in terms of the two normally distributed random values are presented schematically in Figure 4.12.

a) Failure surface is concave: linearization at the "hot point" overestimates the probability of failure.
b) Failure surface is rectilinear: linearization at the "hot point" doesn't produce inaccuracy, since it is not needed.
c) Failure surface is convex: linearization at the "hot point" underestimates the probability of failure.

Let's consider now the most frequent case when the failure surface is indeterminate. Let the boundary of failure region in the standardized space of two random variables in the first coordinate quarter coincide with the circle whose center is at the coordinate origin (Figure 3.13).

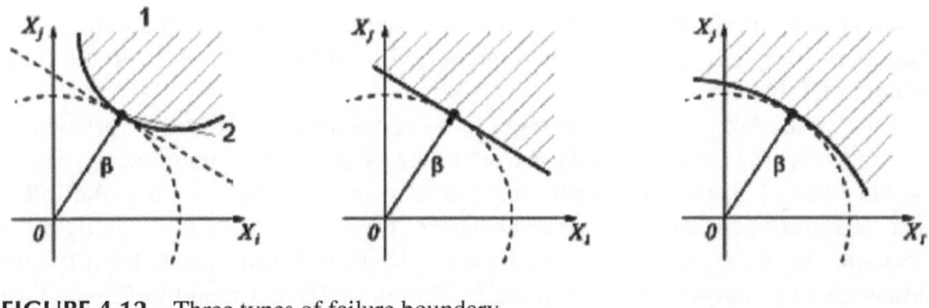

FIGURE 4.12 Three types of failure boundary

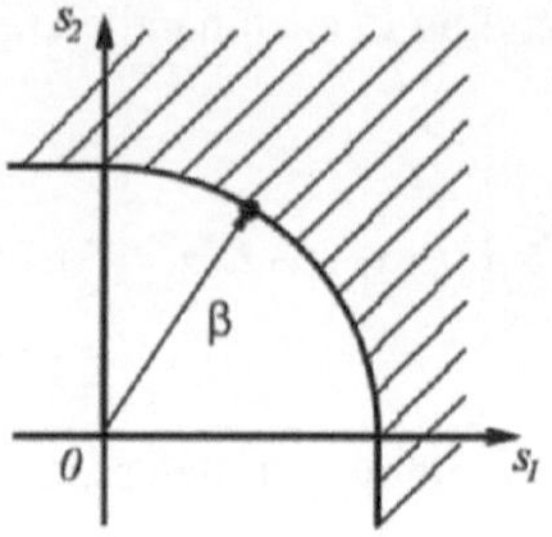

FIGURE 4.13 Diagram of boundary

TABLE 4.4
Dependence of failure probability P_f on reliability index β

	Probability of failure, P_f		
Reliability index, β	**HPM**	**DIDF**	**DIDF/HPM**
2.	0.0228	0.0566	2.48
2.5	0.0066	0.0172	2.61
3.0	0.00135	0.004132	3.06
3.5	0.000233	0.000781	3.35
4.0	$3.17 \cdot 10^{-5}$	0.000116	3.66
4.5	$3.40 \cdot 10^{-6}$	$1.34 \cdot 10^{-5}$	3.94
5.0	$2.87 \cdot 10^{-7}$	$1.22 \cdot 10^{-6}$	4.25
5.5	$1.90 \cdot 10^{-8}$	$8.67 \cdot 10^{-8}$	4.56
6.0	$9.89 \cdot 10^{-10}$	$4.81 \cdot 10^{-9}$	4.86

Table 4.4 shows the relation between failure probability P_f and reliability index β, obtained on the bases of the "hot point" method (HPM) and by direct integration of distribution function (DIDF).

Table 4.5 and Figure 4.14 show the relation between probability of failure P_f and boundary curvature $1/\rho$ for reliability index $\beta = 3.5$ received using the method of direct integration of distribution function. The ratio of the exact solution over that obtained by "hot point" method, namely, $P_f = 0.000233$ is given in the third column of Table 4.5.

The results show that if the values of failure boundary curvature $1/\rho$ vary within the range [0; 0.1] (that is usually a wide range of structures), then inaccuracy of the "hot point" method is rather small. It also makes sense to show the influence of failure boundary curvature on failure probability P_f in the vicinity or at a distance from the "hot point," see Figure 4.15. Calculation results for different values of a and curvature radius ρ are presented in Table 4.6 and in Figure 4.16 for $\beta = 3$.

TABLE 4.5
Failure probability vs. boundary curvature

Curvature, $1/\rho$	Failure probability P_f	$P_{f,exact}$ HPM
0.286	0.000781	3.35
0.25	0.000615	2.64
0.2	0.000458	1.97
0.143	0.000350	1.50
0.1	0.000291	1.25
0.067	0.000271	1.16
0.005	0.000260	1.12
0.033	0.000250	1.07
0.02	0.000243	1.04
0.01	0.000238	1.02
0.001	0.000234	1.004

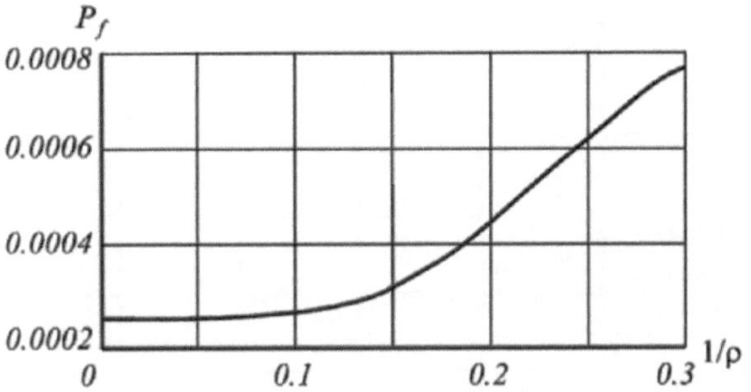

FIGURE 4.14 Dependence of P_f on curvature $1/\rho$

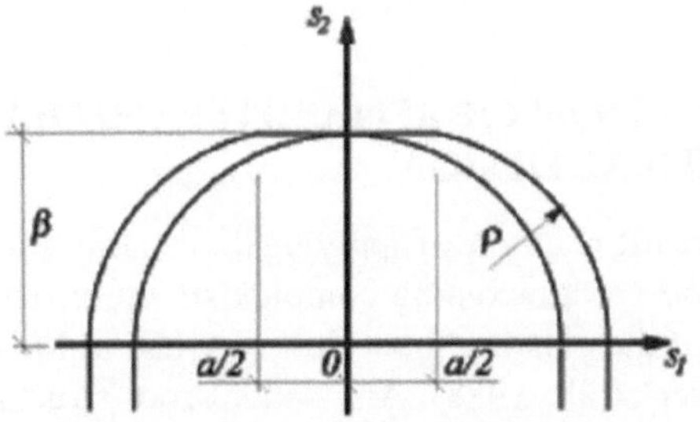

FIGURE 4.15 Vicinity of the "hot point"

TABLE 4.6
Failure probability vs. rectilinear area extent a and failure boundary curvature ρ

Length *a*	Probability of failure P_f				
	$\rho = 3$	$\rho = 4$	$\rho = 5$	$\rho = 8$	$\rho = 13$
0	0.00620	0.00287	0.00222	0.00174	0.00156
1	0.00550	0.00264	0.00208	0.00167	0.00152
2	0.00403	0.00216	0.00180	0.00155	0.00145
3	0.00267	0.00174	0.00156	0.00144	0.00140
4	0.00185	0.00150	0.00143	0.00138	0.00137
5	0.00150	0.00139	0.00137	0.00136	0.00135
6	0.00138	0.00136	0.00136	0.00135	0.00135

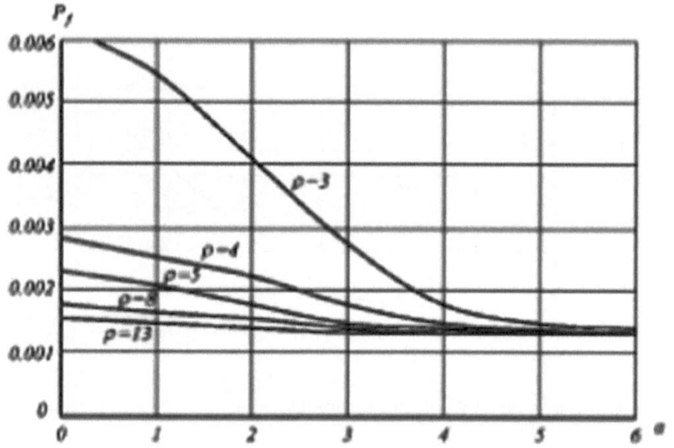

FIGURE 4.16 Failure probability P_f vs. failure boundary ρ (β = 3)

4.7 IMPLEMENTATION OF RELIABILITY THEORY IN STRUCTURAL DESIGN

This section presents a life example of applying the reliability assessment techniques discussed in Chapter 3 for an existing complex structure. This is a multi-element steel spherical roof of the existing giant Sports Arena in Moscow (Russia) shown in Figure 4.17 (Raizer et al., 1988). Any spatial roof structure built over a large congregation area consists of a great number of elements and there is considerable probability of occurring one or more weak elements. Nevertheless, the structure should have a high degree of reliability to ensure human safety. The steel roof structure consists of 1900 elements. It is necessary to analyze the possibility of

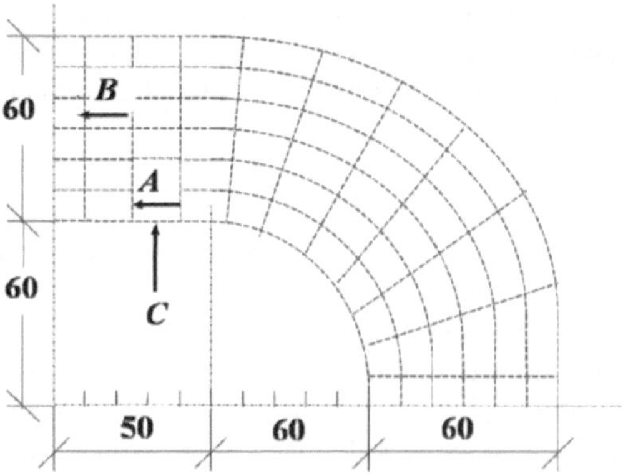

FIGURE 4.17 Layout of elements for a quarter of roof

finding weak elements within a zone of dangerous stress. So, in addition to complying with common design codes, a probabilistic analysis of reliability should be performed considering typical peculiarities of the special technology. The first step was to define the "failure."

Following the human safety requirements, significant plastic deformations should be considered a highly undesirable phenomenon. As "significant" is a too vague definition, using the probability of remaining in the elastic stage appears to be a reasonable measure of reliability. In other words, the failure can be defined as the first appearance of a critical stress condition in any point of the structure. However, one should keep in mind that this measure is only a lower bound to the actual probability of survival because the event is not considered a real failure but just a warning and there is no danger in it itself. The probability of the elastic stage remaining in the structure for n years is:

$$P_{sn} = \int_{-\infty}^{\infty} P_{Fn}(F)\left[1 - P_R(F)\right]dF \tag{4.52}$$

where P_{Fn} is the probability density function (PDF) of the maximum load effect F for n years; P_R is the cumulative distribution function (CDF) of the load bearing capacity R of the whole structure. Assuming conservatively the independence of the yield stresses of the elements,

$$P_R(F) = 1 - \prod_{i=1}^{m}\left[1 - P_{\sigma_y}\sigma_i(F)\right] \tag{4.53}$$

where P_{σ_y} is the CDF of the yield stress σ_y, m is the number of elements, σ_i is the maximum stress in the ith element. One can easily see that at very big number of elements, the second term in (3.53) obviously tends to 0 even at small values of P_{σ_T}. Therefore, if P_{σ_T} is an approximation of the empirical CDF by any unbounded probability law, then the region of values F, where $0 < P_R < 1.0 - \xi$ (ξ is an infinitesimal), will be determined by low-probable very small values of σ_y hardly if ever observed in reality. It leads to an absurdly small value of reliability measure (4.52) of multi-element structure, whereas it is not a defect of structure but a defect of calculation technique. The problem can be solved by using a reasonably bounded empirical CDF for P_{σ_y} in (4.53), which is the outer envelope of test results. A theoretical probability law should be used for calculating the probability P_{sn} (4.46) as an approximation of the P_R CDF. Moreover, it should be a local approximation in the region of maximum values of the integrand (4.53). A structural analysis was performed taking into consideration two types of loads: permanent load (weight of the structure and the roof) and snow load. According to the code of practice, the load effects in elements were calculated as a result of acting the loads with their design values. For the reliability analysis the permanent load was considered to be deterministic equal to the design value, so that F in (4.46) is the random value of snow load and P_{Fn} is the PDF of its maxima for n years. Assuming the linear behavior of the structure within the elastic stage, stress σ_i in (3.53) can be determined as

$$\sigma_i(F) = \sigma_{pi}^d + \sigma_{si}^d \frac{F}{F_d} \tag{4.54}$$

where $\sigma_{pi}{}^d$ and $\sigma_{si}{}^d$ are the stresses in the ith element due to the design values of permanent load and snow load, respectively, and F_d is the design value of snow load.

The technique used here is a variant of the hot point method with a local approximation of the distributions of basic variables by the normal law. In this case, the approximation is carried out at the hot point, which is the point on the failure region boundary in the basic variables space where the approximated joint probability density function achieves its maximum. The hot point can be found out by iterations as illustrated in Figure 4.2.

The probability calculation procedure consists of the following operations. The boundary of the failure region is the line,

$$F = R \tag{4.55}$$

In the first iteration, the "approximation point" (F_{ap}, R_{ap}) is chosen arbitrarily. As it is situated on the failure boundary, then,

$$F_{ap} = R_{ap} \tag{4.56}$$

The distributions of basic variables at point (F_{ap}, R_{ap}) are approximated by the normal distribution that gives $\Phi_F(F_{ap}) = P_{Fn}(F_{ap})$; $\varphi_F(F_{ap}) = p_{Fn}(F_{ap})$; $\Phi_R(F_{ap}) = P_R(F_{ap})$; $\varphi_R(F_{ap}) = p_R(F_{ap})$; where p_{Fn} and p_R are the PDF of the initial distributions; Φ_F, Φ_R, φ_F, and φ_R are the CDF and PDF of the normal distributions. Their parameters are determined by the following equations:

mean values:

$$\mu_R = R_{ap} + u_R s_R \quad \text{and} \quad \mu_F = F_{ap} - u_F s_F \tag{4.57}$$

standard deviations:

$$s_R = \frac{\varphi_R(u_R)}{P_R(F_{ap})} \quad \text{and} \quad s_F = \frac{\varphi_F(u_F)}{P_F(F_{ap})} \tag{4.58}$$

where $u_R = \Phi^{-1}[P_R(F_{ap})]$; $u_F = \Phi^{-1}[P_{Fn}(F_{ap})]$;$\Phi$ and φ are the CDF and PDF of the standardized normal distribution. The distance from the failure boundary to the origin in a space of standardized normal variables is

$$\rho = \frac{u_F s_F - u_R s_R}{\sqrt{s_F^2 + s_R^2}} \tag{4.59}$$

and the distance from the "approximation point" to the origin in the same space is

$$r = \sqrt{u_F^2 + u_R^2} \tag{4.60}$$

If $\rho = r$, then the chosen approximation point is the hot point and the distance of failure is reliability index $\beta = \rho$, that is, $P_{sn} = \Phi(\beta)$. If not, then go to the next iteration assuming

$$F_{ap} = \mu_F + \frac{s_F^2}{\sqrt{s_F^2 + s_R^2}} \rho \tag{4.61}$$

The distribution of snow load was assumed to be the Fisher-Tippet Extreme I distribution with CDF

$$P_F(F) = \exp\left[-\exp\frac{(F-a)}{b}\right] \tag{4.62}$$

where $a = 0.91$ kPa and $b = 0.29$ kPa were determined based on data recorded at the meteorological stations. The distribution of the maxima for n years in (4.52) is

$$P_{Fn}(F) = \exp\left[-\exp\frac{(F - a_n)}{b}\right] \quad (4.63)$$

where $a_n = a + b\ln n$. The values of CDF $P_R(F)$ of the load-bearing capacity (4.53) were calculated for discrete points with a step $\Delta f = \Delta F/\Delta F_d = 0.1$ (Figure 4.18). The set of these points was used as the possible approximation and hot points. The design value of snow load according to the code is $F_d = 1.4$ kPa. The CDF $P_{\sigma y}$ of yield stress σ_y was established as the outer envelope of 158 experimental points, which are the result of testing the specimens of steel used in the structure. The lowest test result was $\sigma_y = 390$ *MPa* and the empirical CDF $P_{\sigma y}$ was assumed bounded at the left by the value $\sigma_y = 370$ *MPa*. As a consequence of this restriction, the CDF $P_R(F)$ proves to be bounded at the left by the value of $F = 3.64$ kPa. Each $P_{\sigma y}[\sigma_i(F)]$ in Equation (3.53) is the CDF of load bearing capacity of the ith element, that is, $P_{Ri}(F) = P_{\sigma y}[\sigma_i(F)]$. It gives an opportunity to analyze the contribution of each element to the reliability of the whole structure.

In-plane view the structure Figure 4.17 (dimensions are in meters) is a ring formed by two separate semicircular parts with straight insertions in between. It consists of sloping radial ribs and horizontal rings. The most vulnerable elements are situated at the straight insertions. Figure 4.18 shows the CDF $P_R(F)$ of the whole structure and the CDF $P_{Ri}(F)$ of the three groups of elements denoted by letters A, B and C (see Figure 4.17) in the order of their vulnerability. Each group consists of four identical elements (one in each quarter of the structure) as the most dangerous case of loading is a uniform snow over the whole roof. It is obvious that reliability of the whole structure is conditioned by the CDF of four elements A. In

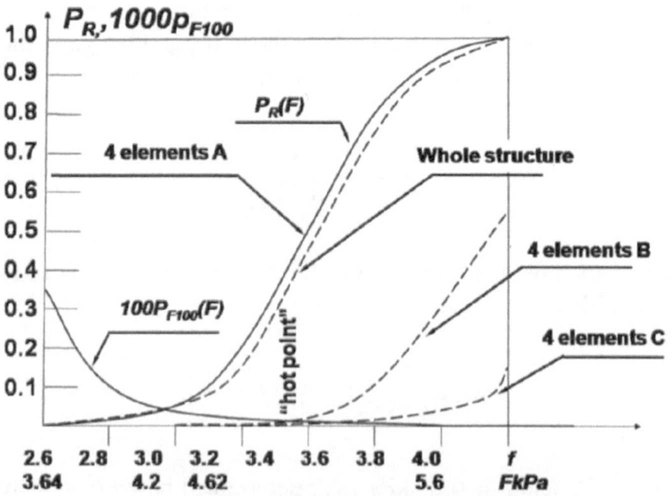

FIGURE 4.18 CDF of load bearing capacity $P_R(F)$ and PDF of snow load $P_{F100}(F)$

other words, if one needs to increase the reliability of the structure, four elements A should to be strengthened first.

For the 100-year service life the hot point was found to be the point with $F = 4.62$kPa and the reliability index $\beta = 3.43$, that is, sought for reliability $P_{s100} = \Phi(3.43) = 0.9997$. The obtained reliability measure is very high, especially as it is the probability of elastic stage. The main aim of this work was to develop a procedure for checking the reliability of multi-element roof structures analyzing the vulnerability of its different parts and elements. Software development allows changing the approach in probabilistic design and abandoning traditional methods of structural mechanics. Modern software makes it possible to perform complex computations for calculating failure probabilities using the above-mentioned or other numerical methods. Their most valuable characteristics are simplicity, no limitations on the nature of initial statistical information, as well as the possibility of using real data (loads, experimental values of physical and geometrical parameters of the structure, etc.) without making preliminary theoretical probabilistic models of stochastic parameters. All of this creates the great opportunity of using probabilistic methods in design.

5 Alternative Definitions of the Fuzzy Safety Factor

> The main action of fuzzy logic is to extend formal considerations from exact world to real world, i.e., fuzzy logic makes it possible to take concepts, notions etc. which are not well-defined, into the formal consideration. Every now and then somebody says that developing fuzzy systems and methods is revolutionary.
>
> Mattila, 2004

> The machinery of linguistic variables and fuzzy if-then rules in fuzzy logic can be employed to improve the performance of genetic algorithms and, more particularly, make it possible to use granulated representations of fitness functions in cases where they are not well defined.
>
> Zadeh, 1998

The concept of safety factor is nearly universally applied in structural engineering. There are presently several probabilistic approaches to the safety factors' interpretation in the literature. Herein we generalize the above approaches to derive four possible definitions of safety factors that are based on fuzzy sets. A fundamental problem of the strength materials, namely an element subjected to a tensile stress, is analyzed in the fuzzy sets' context. Triangular membership function is considered for both stress and strength. Additionally, a structure possessing two failure modes is considered.

5.1 INTRODUCTORY COMMENTS

This section is devoted to the concept of safety factors that are universally utilized in many branches of engineering. In this context, it is instructive to quote Bruhn (1975):

> Most current … vehicles use the Safety Factor approach in the design of structural component. This factor is designed to arbitrarily account for items such as material property variations; manufacturing differences, since no two parts can be made exactly the same; uncertainties in the loading environment; and unknowns in the internal load and stress distributions.

However, as Norton (2000) writes,

> choosing the safety factor is often a confusing proposition for the beginning designer. The safety factor can be thought of as a measure of the designer's

DOI: 10.1201/9781003265993-5

> uncertainty in the analytical models, failure theories, and material property data used, and should be chosen accordingly … Nothing is absolute in engineering any more than in any other endeavor. The strength of materials may vary from sample to sample. The actual size of different examples of the "same" parts made in quantity may vary due to manufacturing tolerance. As a result, we should take the statistical distributions into account in our calculations.

As we see, even the manner of explanation of the safety factor concept is quite critical, for it utilizes the adjectives words like "arbitrary" (Bolotin) and "confusing" (Norton, 2000). Indeed, deterministic approaches to the safety factor assignment were criticized by many authors. Bolotin (1969), for example, stressed that

> the values of safety factors, as well as closely associated values of design loads and design resistances, were improved and modified mainly empirically, by the way of generalization of multi-year experience and exploitation of the structures. Yet, as is seen of the essence of the problem in principle, there are also theoretical approaches possible with wide application of the apparatus of theory of probability and mathematical statistics.

Freudenthal (1957) and Rzhanitsyn (1947) pioneered an explanation of the safety factor in probabilistic context and introduced the so called *central safety factor*. It was shown much later (Elishakoff, 2004) that the probabilistic mechanics offers four possible interpretations of its safety factors. To set the stage for the fuzzy safety factors it appears instructive to describe the probabilistic safety factors first.

5.2 SAFETY FACTOR IN THE PROBABILISTIC FRAMEWORK

Several probabilistic approaches to safety factor have been presented in the literature. Freudenthal (1957) apparently was the first one to introduce the central safety factor defined as the ratio:

$$s_1^{(p)} = \frac{\mathrm{E}(\Sigma_y)}{\mathrm{E}(\Sigma)} \tag{5.1}$$

where the stress Σ and the yield stress Σ_y are treated as random variables and are denoted by capital letters; for their possible realizations σ and σ_y, respectively, the lower-case letters are used; E(·) means mathematical expectation, the superscript "*p*" indicates that we deal with probabilistic case. Birger (1970), and later Maymon (2000) and Qu and Haftka (2002), utilized yet another factor; Maymon (2000) called it a stochastic safety factor, introduced as a random variable:

$$S = \Sigma_y / \Sigma \tag{5.2}$$

The probability distribution function of *S* reads,

$$F_S(s) = Prob\left(S = \frac{\Sigma_y}{\Sigma} \le s\right) \tag{5.3}$$

Birger (1970) demanded the structure to be designed so that this function be equal to some pre-selected value p_0:

$$F_S(s = s_2^{(p)}) = Prob\left(\frac{\Sigma_y}{\Sigma} \le s_2^{(p)}\right) = p_0 \tag{5.4}$$

The value of $s = s_2^{(p)}$ that corresponds to the p_0^{th} fractile of the distribution function $F_S(s)$ is declared as the "stochastic safety factor." This implies, for example, then if $p_0 = 0.05$, that in about 95% of the ensemble of the structure, the realizations of the random safety factor will be at least $s_2^{(p)}$. Qu and Haftka (2002) refer to this quantity as a *stochastic sufficiency factor*. Elishakoff (1983) introduced two other safety factors,

$$s_3^{(p)} = \mathrm{E}\left(\frac{\Sigma_y}{\Sigma}\right) \tag{5.5}$$

$$s_4^{(p)} = \mathrm{E}\left(\Sigma_y\right) \cdot \mathrm{E}\left(\frac{1}{\Sigma}\right) \tag{5.6}$$

The value $s_3^{(p)}$ can be dubbed as a mean safety factor, whereas $s_4^{(p)}$ can be called a multiplicative safety factor. The first and the fourth safety factors have one property in common: the contributions of the randomness of stress and yield stress are separated from each other.

Consider a fundamental problem in strength of materials, namely an element subjected to stress Σ, while Σ_y is the material's yield stress (Figure 5.1).

Let both quantities be random variables. The condition of the successful performance is:

$$\Sigma_y > \Sigma \tag{5.7}$$

The desired random event that the yield stress exceeds the stress, that is, that inequality (5.7) takes place, has a probability that is referred to as *reliability R*:

$$R = Prob\left(\Sigma_y > \Sigma\right) = Prob\left(\frac{\Sigma_y}{\Sigma} > 1\right) = Prob\left(S > 1\right) = 1 - F_S\left(s\right) \tag{5.8}$$

Hence, the probability of failure P_f, that is the complement of the reliability

$$P_f = 1 - R \tag{5.9}$$

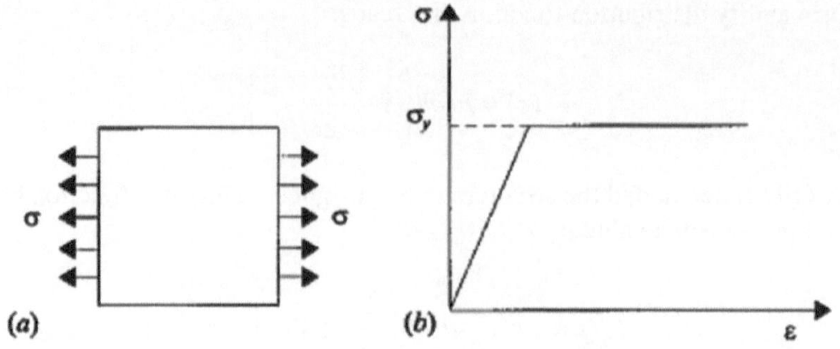

FIGURE 5.1 (a) Element subjected to stress σ; (b) Stress-strain relationship (σ_y = yield stress, = strain)

is obtained as follows:

$$P_f = Prob(S < 1) = F_S(1) \tag{5.10}$$

Thus, the probability of failure equals probability distribution function of the safety factor evaluated at unity.

In this chapter we are interested in introducing the fuzzy sets-based analysis for the safety factor evaluation. This is because we may not possess information that is sufficient to treat Σ and Σ_y as random variables.

5.3 SAFETY FACTOR IN THE FUZZY THEORY FRAMEWORK

In the field of structural engineering, we often face the situation when load and/or strength is imprecise, yet their probabilistic characteristics cannot be obtained. For example, in the case of extreme events, the probabilistic characteristics of these variables cannot always be measured with sufficient accuracy; in such cases (at least sometimes) it appears sensible to treat these variables as fuzzy numbers. Hence, in the design of a structure, one would prefer to define both the stress Σ and the strength Σ_y through their respective membership functions μ_Σ and μ_{Σ_y}. Since Σ and Σ_y are treated as fuzzy numbers, so is their ratio:

$$S = \Sigma_y / \Sigma \tag{5.11}$$

which can be called a *fuzzy safety factor*. In order to find the membership function of the fuzzy safety factor, according to Equation (5.11), an operation of division of two fuzzy numbers Σ_y and Σ must be performed. The membership function μ_S of the fuzzy safety factor can be computed using the so-called α-cuts, that is, for every level of membership α, we consider the α-cut of the stress $\Sigma^\alpha = \{\sigma \in \mathrm{R} | \mu_\Sigma(\sigma) \geq \alpha\} = [\sigma_L^\alpha, \sigma_R^\alpha]$, and α-cut of the yield stress $\Sigma_y^\alpha = \{\sigma_y \in \mathrm{R} | \mu_{\Sigma_y}(\sigma_y) \geq \alpha\} = [\sigma_{Y,L}^\alpha, \sigma_{Y,R}^\alpha]$.

Hereinafter the subscript L will refer to the left branch of the membership function, while the subscript R will be associated with the right branch of it. Thus, the division (Equation (5.11)) becomes an operation between interval variables (Dubois and Prade 1985, Kaufmann and Gupta 1985):

$$S^{\alpha} = [s_L^{\alpha}, \quad s_R^{\alpha}] = \frac{[\sigma_{Y,L}^{\alpha}, \quad \sigma_{Y,R}^{\alpha}]}{[\sigma_L^{\alpha}, \quad \sigma_R^{\alpha}]} = \left[\frac{\sigma_{Y,L}^{\alpha}}{\sigma_R^{\alpha}}, \quad \frac{\sigma_{Y,R}^{\alpha}}{\sigma_L^{\alpha}} \right] \tag{5.12}$$

where S^{α} is the α-cuts of the fuzzy safety factor.

Once the membership function of safety factor μ_S is obtained, defuzzification can be performed in several ways; for example, this can be done via centroid method or by setting some specific value of the α-cut of the membership function. For example, Savoia's (2002) study uses the value $\alpha = 0.05$.

Following Elishakoff (2004) who introduced four probabilistic safety factors, we treat four alternative fuzzy safety factors. We define the *central fuzzy safety factor* as equal to the ratio of the abscissas of the centroids of the membership functions, respectively, of the yield stress and the stress:

$$s_1^{(f)} = G(\Sigma_y) / G(\Sigma) \tag{5.13}$$

where $G(\cdot)$ means the abscissa of center of gravity of an uncertain parameter (fuzzy number), and the superscript "*f*" indicates that we deal with fuzzy numbers. The operator $G(\Sigma)$ is defined as:

$$G(\Sigma) = \frac{\int_{\sigma_L^0}^{\sigma_R^0} \sigma \mu_{\Sigma}(\sigma) \mathrm{d}\sigma}{\int_{\sigma_L^0}^{\sigma_R^0} \mu_{\Sigma}(\sigma) \mathrm{d}\sigma} \tag{5.14}$$

We also define the *characteristic value of fuzzy safety factor* $s_2^{(f)}$ as the lower bound of the α-cut $S^{\alpha} = \left\{ s \in \mathrm{R} | \mu_S(s) \geq \alpha \right\} = [S_L^{\alpha}, S_R^{\alpha}]$ with α specified at the α_0 level, and the latter specified through the design considerations. The characteristic value of fuzzy safety factor $s_2^{(f)}$ is obtained as:

$$s_2^{(f)} = \min\left\{ \ s^{\alpha=\alpha_0} \right\} = \min\left\{ \ [s_L{}^{\alpha=\alpha_0}, \quad s_R{}^{\alpha=\alpha_0}] \ \right\} = s_L{}^{\alpha=\alpha_0} \tag{5.14}$$

We also introduce the *mean value of fuzzy safety factor* $s_3^{(f)}$ as the abscissa of the centroid of the fuzzy safety factor S:

$$s_3^{(f)} = G(S) = G\left(\frac{\Sigma_y}{\Sigma} \right) = \frac{\int_{s_L}^{s_R} s \mu_S(s) \mathrm{d}s}{\int_{s_L}^{s_R} \mu_S(s) \mathrm{d}s} \tag{5.15}$$

It is an analogous to the probabilistic safety factor $s_3^{(p)}$. The "fuzzy" counterpart of the probabilistic safety factor $s_4^{(p)}$ is:

$$s_4^{(f)} = G(\Sigma_y)G\left(\frac{1}{\Sigma}\right) \tag{5.16}$$

Note that if Σ_y is a deterministic quantity while Σ is a fuzzy number, then $s_3^{(f)} = s_4^{(f)}$; if Σ is a deterministic quantity, but Σ_y is a fuzzy number then $s_1^{(f)} = s_3^{(f)} = s_4^{(f)}$.

The feasibility of proposed concepts must be checked to estimate the safety factor in the context of some practical engineering problems. We choose the strength of materials and the stability problems.

5.4 FUNDAMENTAL PROBLEM IN THE STRENGTH OF MATERIALS

In the following an analytical example is reported, with attendant numerical evaluation. Triangular membership functions are employed to represent the fuzzy yield stress Σ_y and the fuzzy actual stress Σ.

Let the stress Σ be a triangular fuzzy variable. The membership function of the stress is represented analytically by Equation (5.18) and is depicted in Figure 5.2:

$$\mu_\Sigma(\sigma) = \begin{cases} 0, & for \;\; \sigma \le \sigma_L^0 \\ \dfrac{\sigma - \sigma_L^0}{\sigma_C - \sigma_L^0}, & for \;\; \sigma_L^0 \le \sigma \le \sigma_C \\ \dfrac{\sigma_R^0 - \sigma}{\sigma_R^0 - \sigma_C}, & for \;\; \sigma_C \le \sigma \le \sigma_R^0 \\ 0, & for \;\; \sigma \ge \sigma_R^0 \end{cases} \tag{5.18}$$

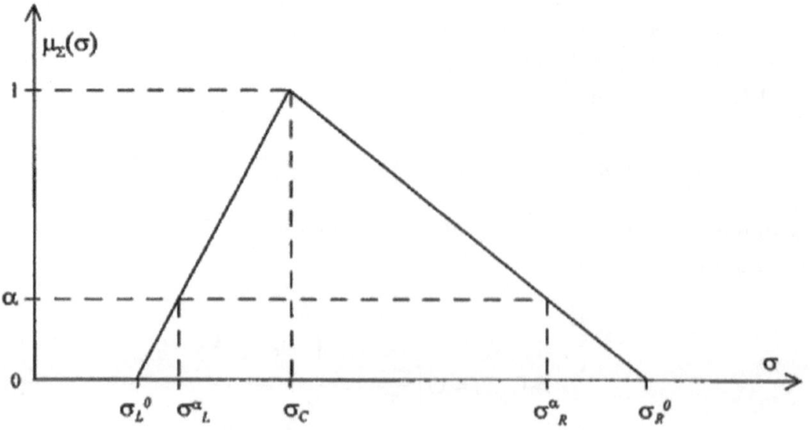

FIGURE 5.2 Membership function of actual stress

The α-cut of the fuzzy stress is:

$$\Sigma^{\alpha} = \left\{ \sigma \in \mathrm{R} | \mu_{\Sigma}(\sigma) \geq \alpha \right\} = [\sigma_L^{\alpha}, \quad \sigma_R^{\alpha}] \tag{5.17}$$

where σ_L^{α} and σ_R^{α} are calculated from Equation (6) by setting $\mu_{\Sigma} = \alpha$. The two branches of membership function are derived:

$$\sigma_L^{\alpha} = \sigma_L^0 + \alpha(\sigma_C - \sigma_L^0), \quad \sigma_R^{\alpha} = \sigma_R^0 - \alpha(\sigma_R^0 - \sigma_C) \tag{5.18}$$

For the sake of simplicity, we also use a triangular membership function of the yield stress Σ_Y:

$$\mu_{\Sigma_Y}(\sigma_Y) = \begin{cases} 0, & \text{for } \sigma_Y \leq \sigma_{Y,L}^0 \\ \dfrac{\sigma_Y - \sigma_{Y,L}^0}{\sigma_{Y,C} - \sigma_{Y,L}^0}, & \text{for } \sigma_{Y,L}^0 \leq \sigma_y \leq \sigma_{Y,C} \\ \dfrac{\sigma_{Y,R}^0 - \sigma_Y}{\sigma_{Y,R}^0 - \sigma_{Y,C}}, & \text{for } \sigma_{Y,C} \leq \sigma_y \leq \sigma_{Y,R}^0 \\ 0, & \text{for } \sigma_Y \geq \sigma_{Y,R}^0 \end{cases} \tag{5.19}$$

The α-cut of yield stress σ_Y reads:

$$\Sigma_Y^{\alpha} = \left\{ \sigma_Y \in \mathrm{IR} \mid \mu_{\sigma_Y}(\sigma_Y) \geq \alpha \right\} = \left[\sigma_{Y,L}^{\alpha}, \sigma_{Y,R}^{\alpha} \right] \tag{5.20}$$

where $\sigma_{Y,L}^{\alpha}$ and $\sigma_{Y,R}^{\alpha}$ are calculated from Equation (5.19) by setting $\mu_{\Sigma_Y} = \alpha$ in the left and right branches of membership function:

$$\sigma_{Y,L}^{\alpha} = \sigma_{Y,L}^0 + \alpha(\sigma_{Y,C} - \sigma_{Y,L}^0), \quad \sigma_{Y,R}^{\alpha} = \sigma_{Y,R}^0 - \alpha(\sigma_{Y,R}^0 - \sigma_{Y,C}) \tag{5.21}$$

From the definition given in Equation (5.11) we calculate the membership function of safety factor S, using Equation (5.12) for each α-cut:

$$S^{\alpha} = [s_L^{\alpha}, \quad s_R^{\alpha}] = \frac{[\sigma_{Y,L}^{\alpha}, \quad \sigma_{Y,R}^{\alpha}]}{[\sigma_L^{\alpha}, \quad \sigma_R^{\alpha}]} = \left[\frac{\sigma_{Y,L}^{\alpha}}{\sigma_R^{\alpha}}, \quad \frac{\sigma_{Y,R}^{\alpha}}{\sigma_L^{\alpha}} \right] \tag{5.22}$$

where

$$s_L^{\alpha} = \frac{\sigma_{Y,L}^{\alpha}}{\sigma_R^{\alpha}} = \frac{\sigma_{Y,L}^0 + \alpha(\sigma_{Y,C} - \sigma_{Y,L}^0)}{\sigma_R^0 - \alpha(\sigma_R^0 - \sigma_C)} \tag{5.23}$$

$$s_R^{\alpha} = \frac{\sigma_{Y,R}^{\alpha}}{\sigma_L^{\alpha}} = \frac{\sigma_{Y,R}^0 - \alpha(\sigma_{Y,R}^0 - \sigma_{Y,C})}{\sigma_L^0 + \alpha(\sigma_C - \sigma_L^0)} \tag{5.24}$$

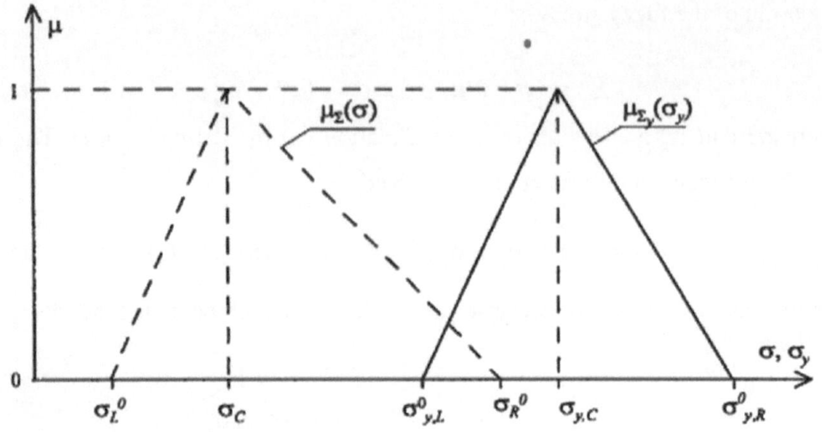

FIGURE 5.3 Triangular membership functions for actual stress and yield stress, respectively

By setting $\alpha = \mu_S$ and $s_L^\alpha = s$, in Equation (5.23) we find the expression for the left branch of the membership function of the safety factor; likewise, by fixing $\alpha = \mu_S$ and $s_R^\alpha = s$ in Equation (5.26) we arrive at the right branch of the membership function of the safety factor. To sum up, μ_S becomes

$$\mu_S(s) = \frac{\Sigma_y}{\Sigma} = \begin{cases} 0, & \text{for } s \le \dfrac{\sigma^0_{Y,L}}{\sigma^0_R} \\ \dfrac{\sigma^0_{Y,L} - \sigma^0_R s}{\sigma^0_{Y,L} - \sigma_{Y,C} + s\,(\sigma_C - \sigma^0_R)}, & \text{for } \dfrac{\sigma^0_{Y,L}}{\sigma^0_R} \le s \le \dfrac{\sigma_{Y,C}}{\sigma_C} \\ \dfrac{\sigma^0_{Y,R} - \sigma^0_L s}{\sigma^0_{Y,R} - \sigma_{Y,C} + s\,(\sigma_C - \sigma^0_L)}, & \text{for } \dfrac{\sigma_{Y,C}}{\sigma_C} \le s \le \dfrac{\sigma^0_{Y,R}}{\sigma^0_L} \\ 0, & \text{for } s \ge \dfrac{\sigma^0_{Y,R}}{\sigma^0_L} \end{cases} \tag{5.25}$$

In order to evaluate the four fuzzy safety factors as per definitions in Equations (5.13), (5.14), (5.15) and (5.16) we have to calculate the abscissas of centroids of the membership function of the yield stress, of the membership function of the actual stress, as well as of the membership function safety factor and of the inverse of stress. The abscissas of centroids of yield stress and actual stress, respectively, read

$$G(\Sigma_Y) = \frac{\sigma_{Y,C} + \sigma^0_{Y,R} + \sigma^0_{Y,L}}{3} \tag{5.26}$$

$$G(\Sigma) = \frac{\sigma_C + \sigma_R^0 + \sigma_L^0}{3} \tag{5.27}$$

Once they are substituted into Equation (5.13), that defines the *first fuzzy safety factor* for the particular case of triangular membership function for stress and yield stress, we obtain,

$$s_1^{(f)} = \frac{G(\Sigma_y)}{G(\Sigma)} = \frac{\sigma_{Y,C} + \sigma_{Y,R}^0 + \sigma_{Y,L}^0}{\sigma_C + \sigma_R^0 + \sigma_L^0} \tag{5.28}$$

From the definition of the second fuzzy safety factor (Equation (5.14)) by setting $\alpha = \alpha_0$ in Equation (5.24) we obtain the *characteristic value of fuzzy safety factor* $s_2^{(f)}$ for the particular case of triangular membership function for stress and yield stress:

$$s_L^{\alpha} = \frac{\sigma_{Y,L}^{\alpha}}{\sigma_R^{\alpha}} = \frac{\sigma_{Y,L}^0 + \alpha(\sigma_{Y,C} - \sigma_{Y,L}^0)}{\sigma_R^0 - \alpha(\sigma_R^0 - \sigma_C)}$$

$$s_2^{(f)} = \min\left\{s^{\alpha_0}\right\} = s_L{}^{\alpha_0} = \frac{\sigma_{Y,L}^0 + \alpha(\sigma_{Y,C} - \sigma_{Y,L}^0)}{\sigma_R^0 - \alpha(\sigma_R^0 - \sigma_C)} \tag{5.29}$$

In the reliability analysis via fuzzy sets, while not directly dealing with the concept of the safety factor, Savoia (2002) utilized the value $\alpha_0 = 0.05$. Following Savoia, we utilize this particular value for the safety factor calculation. We get,

$$s_2^{(f)} = \min\left\{\, s^{\alpha_0} \right\} = s_L{}^{\alpha_0} = \frac{\sigma_{Y,L}^0 + 0.05(\sigma_{Y,C} - \sigma_{Y,L}^0)}{\sigma_R^0 - 0.05(\sigma_R^0 - \sigma_C)} \tag{5.30}$$

The *central fuzzy safety factor* via Equation (5.15) is

$$s_3^{(f)} = G(S) = G\left(\frac{\Sigma_y}{\Sigma}\right) = \frac{\int_{s_L}^{s_R} s\mu_S(s)\mathrm{d}s}{\int_{s_L}^{s_R} \mu_S(s)\mathrm{d}s} \tag{5.31}$$

The analytical expression for $s_3^{(f)}$ is cumbersome. For its calculation, symbolic package was utilized. The formula is reported in the Appendix, to be able to compare its results with other fuzzy safety factors when either Σ or Σ_y turns out to be a deterministic quantity.

Before evaluating the fourth safety factor (Equation (6.15)) we introduce a new fuzzy number T that is reciprocal to Σ:

$$T = 1/\Sigma \tag{5.32}$$

Its α-cut reads:

$$T^{\alpha} = \left(\frac{1}{\Sigma}\right)^{\alpha} = \left[t_L^{\alpha},\ t_R^{\alpha}\right] = \frac{[1,\ 1]}{\left[\sigma_L^{\alpha},\ \sigma_R^{\alpha}\right]} = \left[\frac{1}{\sigma_R^{\alpha}},\ \frac{1}{\sigma_L^{\alpha}}\right] \tag{5.33}$$

where the bounds of the interval T^{α} are:

$$t_L^{\alpha} = \frac{1}{\sigma_R^{\alpha}} = \frac{1}{\alpha(\sigma_R^0 - \sigma_C) + \sigma_R^0} \qquad t_R^{\alpha} = \frac{1}{\sigma_L^{\alpha}} = \frac{1}{\alpha(\sigma_C - \sigma_L^0) + \sigma_L^0} \tag{5.34}$$

By setting $\alpha = \mu_T$ and $t_L^{\alpha} = t$, in the first Equation (5.34) we find the expression for the left branch of the membership function of the inverse T of the stress; likewise, by fixing $\alpha = \mu_T$ and $t_R^{\alpha} = t$ in the second Equation (5.34) we arrive at the right branch of the membership function of T. To sum up, μ_T becomes

$$\mu_T(t) = \mu_{1/\Sigma}\left(\frac{1}{\Sigma}\right) = \begin{cases} 0, & for\ \ t \le \dfrac{1}{\sigma_R^0} \\ \dfrac{1 - \sigma_R^0 t}{t(\sigma_C - \sigma_R^0)}, & for\ \ \dfrac{1}{\sigma_R^0} \le t \le \dfrac{1}{\sigma_C} \\ \dfrac{1 - \sigma_L^0 t}{t(\sigma_C - \sigma_L^0)}, & for\ \ \dfrac{1}{\sigma_C} \le t \le \dfrac{1}{\sigma_L^0} \\ 0, & for\ \ t \ge \dfrac{1}{\sigma_L^0} \end{cases} \tag{5.35}$$

By evaluating the abscissa of centroid of T:

$$G(T) = \frac{\displaystyle\int_{1/\sigma_L^0}^{1/\sigma_R^0} t\ \mu_T(t)\,\mathrm{d}t}{\displaystyle\int_{1/\sigma_L^0}^{1/\sigma_R^0} \mu_T(t)\,\mathrm{d}t} \tag{5.36}$$

we get

$$G(T) = \frac{1}{2}\frac{(\sigma_R^0 - \sigma_L^0)(\sigma_R^0 - \sigma_C)(\sigma_L^0 - \sigma_C)}{\sigma_C\sigma_R^0\sigma_L^0(\sigma_L^0\ln(\sigma_C) + \sigma_C\ln(\sigma_R^0) - \sigma_L^0\ln(\sigma_R^0) - \sigma_C\ln(\sigma_L^0) + \sigma_R^0\ln(\sigma_L^0) - \sigma_R^0\ln(\sigma_C))} \tag{5.37}$$

Using the expression for centroid of the membership function of yield stress (Equation (5.26)) and the centroid of the membership function of inverse of the stress, we derive the analytical expression of the fourth safety factor:

$$s_4^{(f)} = G(\Sigma_y)G(T) =$$

$$= \frac{1}{6}\frac{(\sigma_{Y,C} + \sigma_{Y,R}^0 + \sigma_{Y,L}^0)(\sigma_R^0 - \sigma_L^0)(\sigma_R^0 - \sigma_C)(\sigma_L^0 - \sigma_C)}{\sigma_C\sigma_R^0\sigma_L^0[\sigma_L^0 \ln(\sigma_C) + \sigma_C \ln(\sigma_R^0) - \sigma_L^0 \ln(\sigma_R^0) - \sigma_C \ln(\sigma_L^0) + \sigma_R^0 \ln(\sigma_L^0) - \sigma_R^0 \ln(\sigma_C)]} \tag{5.38}$$

5.5 NUMERICAL EXAMPLES

The numerical values adopted for the fuzzy stress variable are:

$\sigma_L^0 = 120\ MPa \quad \sigma_R^0 = 230\ MPa,\ \sigma_C = 150\ MPa$

For the yield stress we choose as the central value $\sigma_{Y,C}$, the value of yield stress of steel Fe 360 via the Italian code, $\sigma_{Y,C} = 235\ MPa$; the lower and upper bounds of the yield stress are fixed at $\sigma_{Y,L}^0 = 200\ MPa$, $\sigma_{Y,R}^0 = 280\ MPa$. It is noted that $\sigma_{Y,C}$ is greater than σ_C so that with respect to these values the naive "deterministic" safety factor $s = \sigma_{Y,C}/\sigma_C$ is greater than unity. By substituting these values into Equation (5.25), we determine the membership function of the fuzzy safety factor S:

$$\mu_S(s) = \frac{\Sigma_y}{\Sigma} = \begin{cases} 0, & \text{for } s \le \frac{200}{230} \\ \frac{230s - 200}{80s + 35}, & \text{for } \frac{200}{230} \le s \le \frac{235}{150} \\ \frac{280 - 120s}{45 + 30s}, & \text{for } \frac{235}{150} \le s \le \frac{280}{120} \\ 0, & \text{for } s \ge \frac{280}{120} \end{cases} \tag{5.39}$$

We calculate the four possible safety factors $s_j^{(f)}$ (j = 1, 2, 3, 4). From Equation (5.28) we determine the first safety factor:

$$s_1^{(f)} = \frac{G(\Sigma_y)}{G(\Sigma)} = \frac{\sigma_{Y,C} + \sigma_{Y,R}^0 + \sigma_{Y,L}^0}{\sigma_C + \sigma_R^0 + \sigma_L^0} = 1.43 \tag{5.40}$$

The defuzzified counterpart of the safety factor is taken as the lower bound of the α-cut of the fuzzy safety factor S depicts the evaluation of the second fuzzy safety factor $s_2^{(f)}$ in relation to the chosen value of α_0. It is instructive to pose the following question: What should the α-cut be so the safety factor exceeds unity? Here, for $\alpha_0 \ge \alpha_0^*$ where $\alpha_0^* = 0.295$, the safety factor is greater than unity. Thus, the choice of the α_0 value is very important in order to decide if the system is safe or not. Interestingly, the choice of $\alpha_0 = 0.05$ yields

$$s_2^{(f)} = \min\{s^{\alpha_0}\} = s_L{}^{\alpha_0} = \frac{\sigma_{Y,L}^0 + 0.05(\sigma_{Y,C} - \sigma_{Y,L}^0)}{\sigma_R^0 - 0.05(\sigma_R^0 - \sigma_C)} = 0.8926 \qquad (5.41)$$

The value of α that would lead to the safety factor $s_2^{(f)}$ greater than unity equals $\alpha_0^* = 0.295$. It appears that the "acceptable" values of α-cuts need further study on practical examples, and to calibrate with the current practice.

The third safety factor is the centroid of the membership function of fuzzy safety factor,

$$s_3^{(f)} = G(S) = 1.5512 \qquad (5.42)$$

The fourth safety factor is obtained from the following equation,

$$s_4^{(f)} = G(\Sigma_y)G(1/\Sigma) = 1.5113 \qquad (5.43)$$

In Figure 5.4 the membership function of the fuzzy safety factor S is depicted.

The values of the fuzzy safety factors are $s_1^{(f)} = 1.43$, $s_2^{(f)} = 0.8926$, $s_3^{(f)} = 1.5512$, $s_4^{(f)} = 1.5113$. It was deemed important to present all four possible cases in order for the designers to make a choice on which safety factor is most appropriate in engineering applications.

5.6 STABILITY PROBLEM

In this section, a simple steel structure is considered, consisting of a beam and a column (see Figure 5.4). Note that the constraint of the column presents a flexural

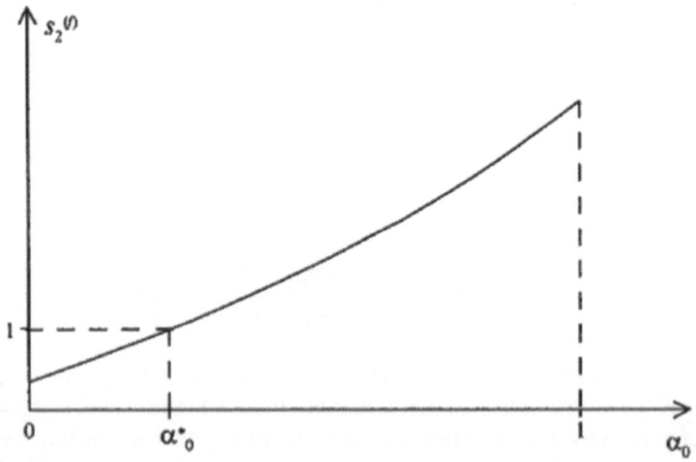

FIGURE 5.4 Relationship between second fuzzy safety factor and membership level

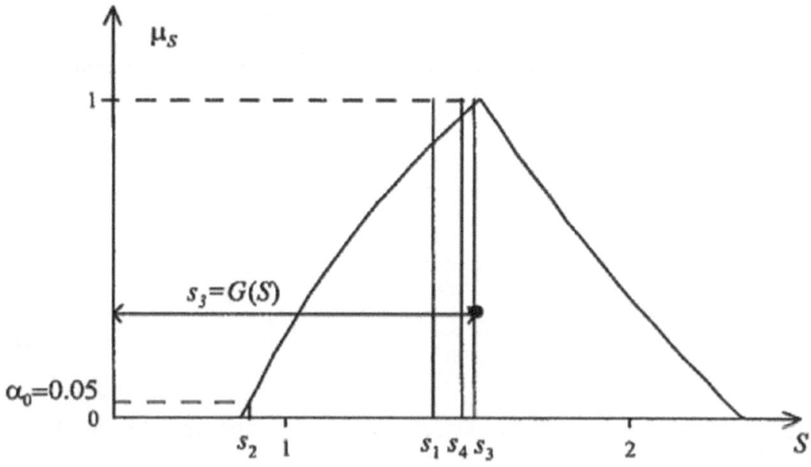

FIGURE 5.5 Membership function of fuzzy safety factor S and four different safety factors for case considered in example

stiffness *k*. Two simple sub-structures representing the free-body diagram are depicted in Figure 5.4. In Figure 5.5 the bending moment diagram is shown for the beam, while in Figure 5.5 the axial force of the column is represented.

This structure has two possible failure modes: bending failure in the beam or instability of the slender column. Hence, in order to check if the structure is safe or not, we have to take into account the two possible failures.

5.7 BENDING FAILURE OF THE BEAM

The maximum bending moment is present in section B, thus the maximum stress is

$$\sigma_{beam} = \frac{M_{\max}}{W_{beam}} = \frac{3}{16}\frac{Fl}{W_{beam}} \tag{5.44}$$

where W_{beam} is section modulus. The beam is safe if the following inequality is fulfilled:

$$s_{beam} = \frac{\sigma_Y}{\sigma_{beam}} \geq 1 \tag{5.45}$$

However, this condition is not sufficient to make sure that the structure is safe because it is possible that another failure mode occurs, namely that the instability of the column takes place.

5.8 STABILITY ANALYSIS OF THE COLUMN

To take in account also stability of the column, the additional inequality has to take place:

$$s_{col} = \frac{\sigma_{cr}}{\sigma_{col}} \geq 1 \tag{5.46}$$

where σ_{cr} is the critical Euler stress, written as

$$\sigma_{cr} = \frac{P_{cr}}{A_{col}} = \frac{\pi^2 EI_{col}}{l_0^2 A_{col}} = \frac{\pi^2 E}{\lambda^2} \tag{5. 47}$$

where h = length, $l_0 = \beta h$ is the effective length, β is a coefficient depending upon the boundary conditions of the element, and λ is the slenderness. The actual stress in the column caused by the reaction X and reads as

$$\sigma_{col} = \frac{X}{A_{col}} = \frac{11}{8}\frac{F}{A_{col}} \tag{5.48}$$

Where A_{col} is the cross-section area of the column. Note that for non-slender columns Equation (5.47) must be substituted by other formula.

In order to be able to declare the structure safe, the two inequalities have to be fulfilled. In Table 5.1 the variables which govern the problem are reported. Until now we have described the problem in deterministic terms; in the next section we consider the case in which load F and the coefficient depending upon the boundary conditions of the element are represented as fuzzy numbers.

5.9 FUZZY SETS-BASED APPROACH

We consider the case in which load F and the coefficient depending upon the boundary conditions are triangular fuzzy variables, with the α-cuts reading, respectively, as

$$F^{\alpha} = \left\{f \in \mathrm{R} \mid \mu_F(f) \geq \alpha\right\} = [f_L^{\alpha}, f_R^{\alpha}] = [f_L^0 + \alpha(f_C - f_L^0),\ \ f_R^0 - \alpha(f_R^0 - f_C)] \tag{5.49}$$

$$\mathrm{B}^{\alpha} = \left\{\beta \in \mathrm{R} \mid \mu_{\mathrm{B}}(\beta) \geq \alpha\right\} = [\beta_L^{\alpha}, \beta_R^{\alpha}] = [\beta_L^0 + \alpha(\beta_C - \beta_L^0),\ \ \beta_R^0 - \alpha(\beta_R^0 - \beta_C)] \tag{5.50}$$

Note that if load F is a fuzzy number the stresses of beam $_{beam}$ and column $_{col}$ are also fuzzy numbers. Likewise, the critical stress $_{cr}$ becomes a fuzzy number, because it is a function of the fuzzy number.

By using the α-cut of load F (Equation (5.49)) we arrive at the α-cut of stresses for beam $_{beam}$ and column $_{col}$, written as follows

$$\Sigma_{beam}^{\alpha} = \left[\frac{3l}{16W_{beam}}, \quad \frac{3l}{16W_{beam}} \right] \cdot [f_L^{\alpha}, \quad f_R^{\alpha}] = [af_L^{\alpha}, \quad af_R^{\alpha}] \tag{5.51}$$

where $a = 3l / 16W_{beam}$, and

$$\Sigma_{col}^{\alpha} = \left[\frac{11}{8A_{col}}, \quad \frac{11}{8A_{col}} \right] \cdot [f_L^{\alpha}, \quad f_R^{\alpha}] = [bf_L^{\alpha}, \quad bf_R^{\alpha}] \tag{5.52}$$

where $b = 11 / 8A_{col}$.

Below we will show how to evaluate the membership function of the critical stress σ_{cr} via the membership function of the length of column μ_K. The result for $\mu_{\Sigma_{cr}}$ reads:

$$\mu_{\Sigma_{cr}}(\sigma_{cr}) = \mu\left(\frac{\pi^2 EI_{col}}{h^2 A_{col}} \frac{1}{\beta^2}\right) = \mu\left(\frac{d^2}{\beta^2}\right) =$$

$$= \begin{cases} 0, & for \ \sigma_{cr} \le \dfrac{d^2}{(\beta_R^0)^2} \\ \dfrac{\beta_R^0}{(\beta_R^0 - \beta_C)} - \dfrac{d}{(\beta_R^0 - \beta_C)\sqrt{\sigma_{cr}}}, & for \ \dfrac{d^2}{(\beta_R^0)^2} \le \sigma_{cr} \le \dfrac{d^2}{(\beta_C)^2} \\ \dfrac{d}{(\beta_C - \beta_L^0)\sqrt{\sigma_{cr}}} - \dfrac{k_L^0}{\beta_C - \beta_L^0}, & for \ \dfrac{d^2}{(\beta_C)^2} \le \sigma_{cr} \le \dfrac{d^2}{(\beta_L^0)^2} \\ 0, & for \ \sigma_{cr} \ge \dfrac{d^2}{(\beta_L^0)^2} \end{cases} \tag{5.53}$$

Moreover, the safety factors S_{beam} and S_{col} themselves are fuzzy numbers:

$$S_{beam} = \frac{\sigma_Y}{\Sigma_{beam}}, S_{col} = \frac{\Sigma_{cr}}{\Sigma_{col}} \tag{5.54}$$

The membership function of S_{beam} reads:

$$\mu_{S_{beam}}(s_{beam}) = \mu\left(\frac{\sigma_Y}{\Sigma_{beam}}\right) = \begin{cases} 0, & for \ s_{beam} \le \dfrac{\sigma_Y}{af_R^0} \\ \dfrac{f_R^0 s_{beam} - \sigma_Y / a}{(f_R^0 - f_C) s_{beam}}, & for \ \dfrac{\sigma_Y}{af_R^0} \le s_{beam} \le \dfrac{\sigma_Y}{af_C} \\ \dfrac{\sigma_Y / a - f_L^0 s_{beam}}{(f_C - f_L^0) s_{beam}}, & for \ \dfrac{\sigma_Y}{af_C} \le s_{beam} \le \dfrac{\sigma_Y}{af_L^0} \\ 0, & for \ s_{beam} \ge \dfrac{\sigma_Y}{af_L^0} \end{cases}$$

The membership function of S_{col} equals:

$$\mu_{S_{col}}\left(s_{col}\right)=\mu\left(\frac{\Sigma_{cr}}{\Sigma_{col}}\right)=$$

$$\Sigma^{\alpha}_{S_{col}}=[\sigma^{\alpha}_{crL},\ \ \sigma^{\alpha}_{crR}]:[\sigma^{\alpha}_{colL},\ \ \sigma^{\alpha}_{colR}]=\left[\frac{\sigma^{\alpha}_{crL}}{\sigma^{\alpha}_{colR}},\ \ \frac{\sigma^{\alpha}_{crR}}{\sigma^{\alpha}_{colL}}\right]$$

6 Convex Models of Uncertainty

It is not implied that this use [of the theory of probability] is in itself sufficient to make a design more reliable or more economical, any more than that the avoidance of the probabilistic approach makes it safer.

A.M. Freudenthal, 1972

In design situations, exact distributional forms are sometimes in doubt.

E.B. Haugen, 1980

I also find that many designers have a "gut feeling" that there is something not quite right with the statistics – the "lies, damn lies, and statistics" school of thought – perhaps not without justification.

A.D.S. Carter, 1997

One of the traditional methods to avoid failure ... is to do a "worst case design." The values of the various variables are taken at the $\pm 3\sigma$ of the dispersion range, and the structure is designed to survive these extreme values.

Giora Maymon, 2002

Good data on load and strength properties are often not available ... in attempting to make account of variability, we are introducing assumptions that might be not tenable, e.g., by extrapolating the load and strength data to the very low probability tails of the assumed population distributions.

P.D.T. O'Connor, 1985

In this chapter we first illustrate some difficulties associated with probabilistic calculations of the reliability. It appears that these and other difficulties are either overlooked or pushed under the rug, as it were, by the probabilistic analysts. Then the alternative approach called "convex modeling" is presented, and the safety factor's associated definition is suggested.

6.1 INTRODUCTORY COMMENTS

One can construct numerous simple examples that demonstrate sensitivity of the failure reliability to the tails of the probability distributions of the random values

DOI: 10.1201/9781003265993-6

involved. It can be directly seen on an elementary example of evaluation of probabilistic characteristics. The example is due to Apostolakis and Kaplan (1965). Consider a cube with a side length X that is a random variable taking on two values, $x_1 = 10^{-2}$ and $x_2 = 2$, with different probabilities:

$$Prob\left(X = 10^{-2}\right) = p_1 = 0.9999 \tag{6.1}$$

$$Prob\left(X = 2\right) = p_2 = 0.0001$$

The mean value of the side length is

$$E(X) = x_1 p_1 + x_2 p_2 = 10^{-2} \times 0.9999 + 2 \times 0.0001 = 1.02 \times 10^{-2} \tag{6.2}$$

The major contribution to this value $E(X) = 1.02\times10^{-2}$ comes from the possible value $x_1 = 10^{-2}$, which has the highest probability. Suppose, now, that we are interested in the mean value of the cube's volume. The volume takes a value $x_1^3 = (10^{-2})^3 = 10^{-6}$ with probability 0.9999; the cube's volume takes on a value $x_2^3 = 8$ with probability 0.0001. The mean volume equals

$$E(X) = x_1 p_1 + x_2 p_2 = 10^{-2} \times 0.9999 + 2 \times 0.0001 = 1.02 \times 10^{-2} \tag{6.3}$$

As is seen, the mean volume is mostly decided by the value $x_2 = 2$ which is taken with a very small probability $p_2 = 10^{-4}$, and not by the value $x_1 = 10^{-2}$ taken on with much higher probability, the factor p_1/p_2 being equal to 9999!

Blekhman, Myshkis and Panovko (1990) write:

> Significantly, the weakness of numerous works on stochastic models-sometimes ruling out any application-lies in the choice of statistical hypothesis, especially of assumptions regarding the probalistic features of the given accidental quantities and functions. These features are often regarded as fully known (like an assumption of a normal distribution will known parameters), or as amenable of determination. In real situations, it mostly turns out that the needed information is lacking.

Harris and Soms (1983) note:

> relatively small perturbations of tail of the strength distribution can make the failure probability far higher than may be desirable, particularly, when failure can be catastrophic.

Yao (1994) writes:

> it has been well known that the failure probability of engineering systems is highly sensitive to the tail portions of relevant distribution functions ... It is not easy for me ... to understand the necessity of computing the failure probability or reliability using various distributions, the tail portions of which remain to be difficult to ascertain.

Ang and Amin (1969) state:

> In reality, of course, the random characteristics of [strength] *R*and [stress] *S* are unknown. Specifically, the form of the distribution functions and associated parameters are unknown; it is possible only to make predictions of what these are, or might be, using appropriate theoretical models. Such theoretical predictions, however, are invariably imperfect and thus subject to errors ...
>
> At high-risk levels, e.g., $P_f \geq 10^{-3}$, differences in the shape of the distribution ... would not sufficiently affect the calculated value of P_f; consequently, the choice of the distribution function for either variable ... would not be too important.
>
> At very low risk, say $P_f \leq 10^{-5}$, the calculated P_f will depend significantly on the distribution [of stress and of strength], so that the failure probability can be determined only with a knowledge of the correct distribution.

In the following section we will demonstrate the high sensitivity of the failure probability; the latter rather than mathematical expectation is of most importance to the application of design of structural components.

6.2 SENSITIVITY OF FAILURE PROBABILITY

Consider an elastic bar compressed by an axial force. The uncertainty parameter is described by the non-vanishing eccentricities e_1 and e_2 for the force (Figure 6.1).

The differential equation describing the deflection of the bar reads:

$$EI\frac{d^4w}{dx^4} + P\frac{d^2w}{dx^2} = 0, \qquad (0 \leq x \leq L) \tag{6.4}$$

where *EI* is the flexural stiffness, *P* the axial force, *w* the displacement, and *L* the length of the bar. Denoting

$$k^2 = \frac{P}{EI} \tag{6.5}$$

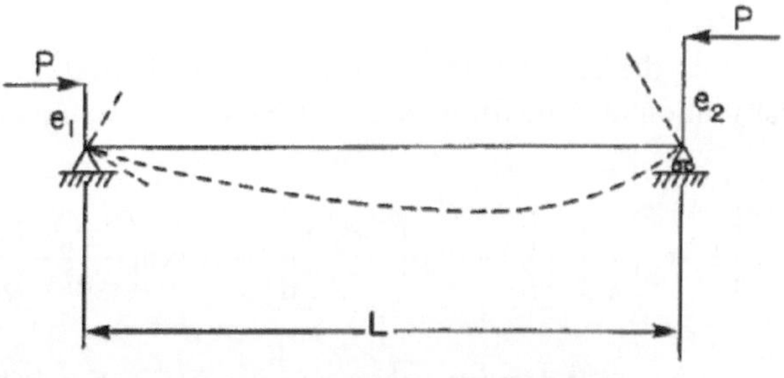

FIGURE 6.1 Beam-column subjected to eccentric forces

The boundary conditions in terms of the bending moment are:

$$M_z(0) = Pe_1, \quad M_z(L) = Pe_2 \tag{6.6}$$

Compliance with the boundary conditions yields the final expression for $M_z(x)$:

$$M_z(x) = -\frac{P}{\sin kL}\left(e_2 - e_1 \cos kL\right)\sin kx + Pe_1 \cos kx \tag{6.7}$$

The maximal bending moment M_z^* is:

$$M_z^*(e_1, e_2) = -\frac{P}{\sin kL}\sqrt{e_1^2 + e_2^2 - 2e_1e_2 \cos kl} \tag{6.8}$$

This expression coincides with Equation 44 in Pikovsky (1961). The problem of a bar in compression with two eccentricities was also studied by Timoshenko and Gere (1963).

One can show that for the maximum bending moment to take place inside the bar, $0<x^*<L$, the following conditions must be satisfied:

$$\cos kL < \frac{e_1}{e_2} < \frac{1}{\cos kL} \tag{6.9}$$

and

$$0 < P < \frac{\pi^2 EI}{4L^2}, \quad 0 < kL < \frac{\pi}{2} \tag{6.10}$$

It can be shown that the maximum bending moment occurs inside the bar and condition (Equation 6.9) is dispensed with, in the following range of load variation:

$$\frac{\pi^2 EI}{4L^2} < P < \frac{\pi^2 EI}{L^2} \tag{6.11}$$

We assume now that the eccentricities constitute a random vector with a jointly exponential distribution and the following distribution function (Gumbel, 1960):

$$F_{E_1E_2}(e_1, e_2) = \left[1 - \exp\left(\frac{e_1}{\beta}\right)\right]\left[1 - \exp\left(-\frac{e_2}{\gamma}\right)\right]\left[1 + \alpha\exp\left(-\frac{e_1}{\beta} - \frac{e_2}{\gamma}\right)\right], \tag{6.12}$$

where e_1 and e_2 take on only positive values and $\beta = E(E_1)$, $Y = E(E_2)$, $E(\bullet)$ denote mathematical expectation.

For the sake of simplicity, we will concentrate on the case presented by Equation (6.6). We are interested in the reliability of the bar, defined as the probability of non-exceedance of a limiting value m^* by the random variable M^*

$$R = \text{Prob}\left(M^* = \frac{P}{\sin kL}\sqrt{E_1^2 + E_2^2 - 2E_1E_2\cos kL} \le m^* \right) \tag{6.13}$$

The integration results in

$$\begin{aligned} F_{M^*}(m^*) = 1 &- \frac{2\beta^3 + \alpha\beta^2\gamma - 5\beta^2\gamma - 2\alpha\beta\gamma^2 + 2\beta\gamma^2}{2\beta^3 - 7\beta^2\gamma + 7\beta\gamma^2 + 2\gamma^3}\exp\left(-\frac{m^*\sin kL}{\beta P}\right) \\ &- \frac{2\alpha\beta^2\gamma - 2\beta^2\gamma - \alpha\beta\gamma^2 + 5\beta\gamma^2 - 2\gamma^3}{2\beta^3 - 7\beta^2\gamma + 7\beta\gamma^2 + 2\gamma^3}\exp\left(-\frac{m^*\sin kL}{\gamma P}\right) \\ &- \frac{\alpha\beta\gamma(2\gamma - \beta)}{2\beta^3 - 7\beta^2\gamma + 7\beta\gamma^2 + 2\gamma^3}\exp\left(-\frac{2m^*\sin kL}{\gamma P}\right) \\ &- \frac{\alpha\beta\gamma(\gamma - 2\beta)}{2\beta^3 - 7\beta^2\gamma + 7\beta\gamma^2 + 2\gamma^3}\exp\left(-\frac{2m^*\sin kL}{\beta P}\right) \end{aligned} \tag{6.14}$$

when following restriction holds

$$2\beta^3 - 7\beta^2\gamma + 7\beta\gamma^2 - 2\gamma^3 \ne 0 \tag{6.15}$$

In the particular cases where instead of the inequality in Equation (6.15), we have an equality, the expressions for the reliability read:

$$\begin{aligned} F_{M^*}(m^*) &= 1 - \left(1 + \frac{m^*\sin kL}{\beta P}\right)\exp\left(-\frac{m^*\sin kL}{\beta P}\right) \\ &-\alpha\left[\left(\frac{m^*\sin kL}{\beta P} - 3\right)\exp\left(-\frac{m^*\sin kL}{\beta P}\right)\right] + \left(\frac{2m^*\sin kL}{\beta P} + 3\right)\exp\left(-\frac{m^*\sin kL}{\beta P}\right), \; for\, \beta = \gamma \end{aligned} \tag{6.16}$$

and

$$\begin{aligned} F_{M^*}(m^*) &= 1 - 2\left(1 - \frac{\alpha}{3}\right)\exp\left(-\frac{m^*\sin kL}{\beta P}\right) \\ &+ \left(1 + \frac{2\alpha m^*\sin kL}{\beta P}\right)\exp\left(-\frac{2m^*\sin kL}{\beta P}\right) + \frac{2\alpha}{3}\exp\left(-\frac{4m^*\sin kL}{\beta P}\right), \; for\, \beta = 2\gamma \end{aligned} \tag{6.17}$$

Finally,

$$F_{M^*}(m^*) = 1 - 2\left(1 - \frac{\alpha}{3}\right)\exp\left(-\frac{m^* \sin kL}{2\beta P}\right) \tag{6.18}$$

$$+\left(1 + \frac{\alpha m^* \sin kL}{\beta P}\right)\exp\left(-\frac{m^* \sin kL}{\beta P}\right) + \frac{2\alpha}{3}\exp\left(-\frac{2m^* \sin kL}{\beta P}\right), for\, \beta = \gamma/2$$

It can be shown (Gumbel, 1960) that the following restrictions should be met, $F_M{}^*(m^*)$ to serve as the distribution function:

$$1 \le \alpha \le 1$$
$$\alpha = 4\rho$$

Also, k could be expressed as

$$k = \frac{\pi}{L}\sqrt{\frac{P}{P_{cl}}} \tag{6.19}$$

where P_{cl} is the classical buckling load of the simply supported bar $P_{cl} = \pi^2 EI/L^2$. The reliability equals

$$R = Prob\left(\Sigma \le \sigma_y\right) = Prob\left[\frac{M^* c}{I} \le \sigma_y\right] = F_{M^*}\left[\frac{\sigma_y I}{c}\right] \tag{6.20}$$

where Σ is the maximum stress, σ_y is yield stress and is assumed to be constant, I is the moment of inertia and c is the distance between the centroidal line and the extreme fiber where the maximum stress occurs. Reliability of the structure is obtainable from Equations (6.14)–(6.18) by replacing m^* by $\sigma_y I/c$.

Figures 6.2–6.4 depict the reliability of the structure versus c.

Figure 6.2 is associated with P = 15 kN, P/P_{cl} = 0.569. The following data is used in Figure 6.6: P = 20 kN, P/P_{cl} = 0.759, whereas in Figure 6.4 the data is fixed at P = 23 kN, P/P_{cl} = 0.873. In all three figures β = 2mm, γ = 1.5 mm, and I/c = 1,333.3 mm^3, L = 1,000 mm, E = 200,000 MPa. As we see, the increase in the applied loading results in reduced reliability of the structure. The coefficient α is varied in Figures 6.2–6.5, in the range $-0.99<\alpha<0.99$. Whereas data on β and γ may be reliable, information on the coefficient a could be insufficient, so that Figures 6.2–6.4 demonstrate the possible scatter in the reliability of the structure due to imprecise knowledge of parameter α. This is illustrated in Figure 6.5, which addresses the design problem: Find the radius of the cross-section c so that the required codified reliability r, or codified probability of failure P_{f^*} = 1-r, will be achieved.

If we fix value of P_{f^*} at 0.01, then, if the calculations are based on α = 0.99, the design value of the radius is c = 12.257; now, if the true value of α is –0.99

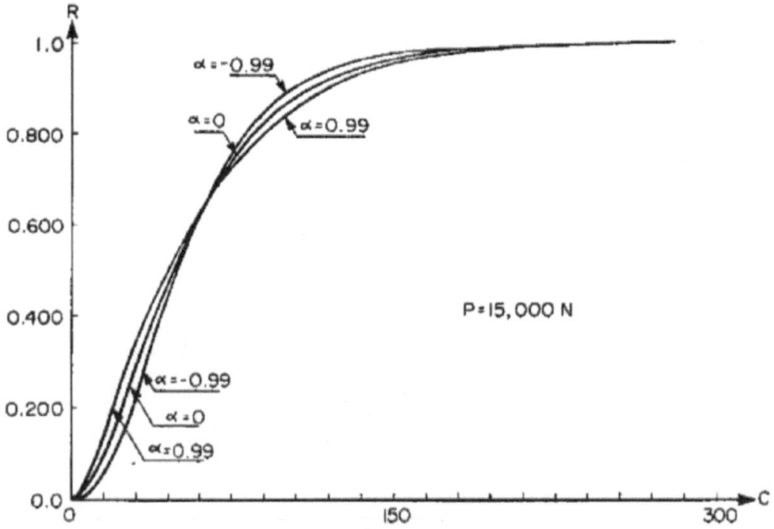

FIGURE 6.2 Structural reliability versus radius

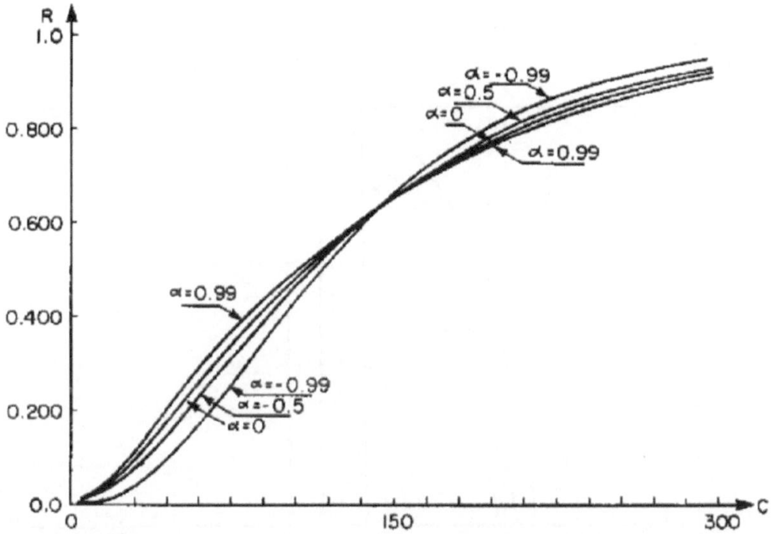

FIGURE 6.3 Dependence of reliability upon parameters

then the actual probability of failure of the system at this radius is 3.74×10^{-3}, that is, it is lower than the codified probability of failure. This implies that we had a case of the "favorable" imprecision. However, if our calculations are based on value $\alpha = -0.99$, then the minimum required radius of the cross-section should be $c = 12.029$. If, however the true value of α is 0.99, then the chosen value of the

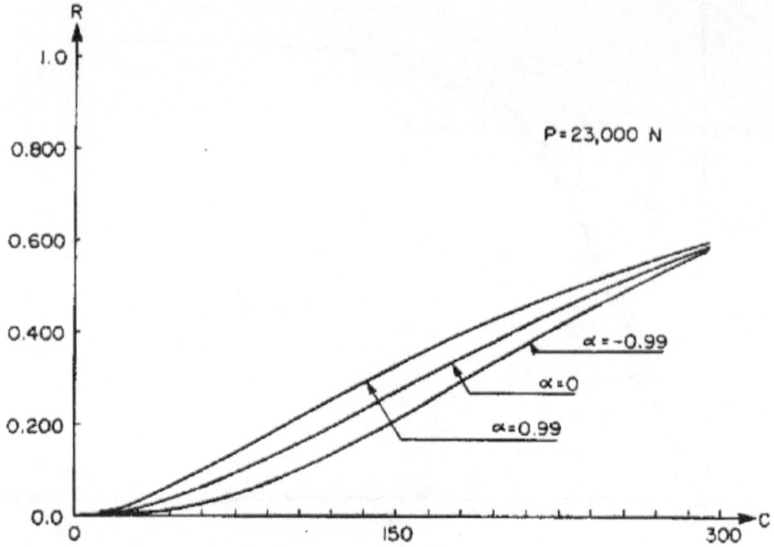

FIGURE 6.4 Influence of the correlation coefficient

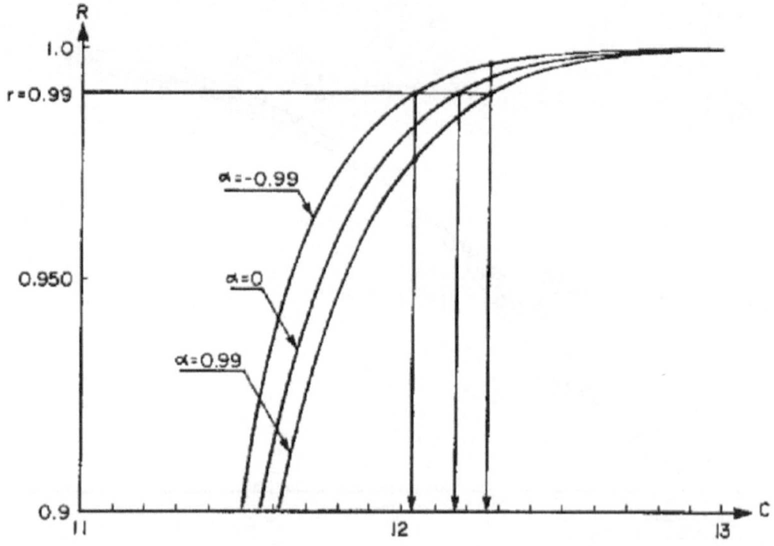

FIGURE 6.5 Solution of the design problem

radius corresponds to actual probability of failure 0.0234 instead of $P_{f*} = 0.01$. This corresponds to the underestimation of the probability of failure by more than twice.

The situation is more severe for highly reliable structures. To get more insight we define the underestimation factor as the ratio of the actual-to-codified probabilities of failure,

$$\eta = \frac{P_f}{P_{f*}} \tag{6.21}$$

For $P_{f*} = 10^{-3}$, the underestimation factor is over 3; for $P_{f*} = 10^{-4}$, $\eta = 3.47$; for $P_{f*} = 10^{-5}$, $\eta = 3.705$; and finally for $P_{f*} = 10^{-6}$ the underestimation factor reaches 3.82.

Thus, one would conclude that the system is acceptable for use, whereas the actual probability of failure is exceeding the codified one, and the system in fact is in a failed state, since the actual reliability is lower than the codified one.

Under these circumstances of the high sensitivity of probability of failure, the natural question arises on how the probabilistic analysis could have been used for design purposes. To attempt to answer this question, we will visualize that the total cost T of production of the column is expressible as

$$T = \frac{q_1}{E(E_1) + E(E_2) + q_2} \tag{6.22}$$

where q_1 and q_2 are constants. Such a postulation maintains that more cost is associated with finer manufacturing, that is, the one with less $E(E_j)$. Figure 6.6 depicts the reliabilities of the columns associated with different mean values of the imperfections $E(E_j)$, while their sum $E(E_1) + E(E_2)$ is kept constant.

Figure 6.6 demonstrates that the maximum reliability is achieved for the equal mean imperfection parameters $E(E_1) = E(E_2)$. Thus, reliability studies could be

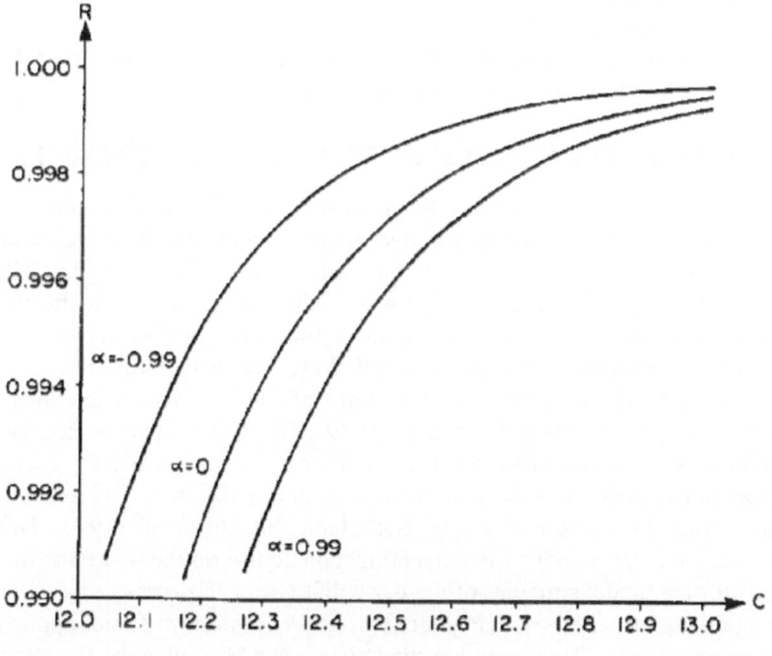

FIGURE 6.6 How does the required high reliability influence the decision making

utilized for comparative purposes; under other conditions being equal, one prefers the manufacturing process, which leads to higher reliability.

To digress, it is instructive to reproduce Jensen's (2000) comments:

> "It mustn't break." This was a sentiment I heard expressed at a recent meeting, where a large corporation had invited one of its major customers to give a presentation and discuss the performance of the manufacturer's products. It was a product, where the competition is very, very fierce, and the revenue earned by the product to function continuously, twenty-four hours a day, every day of the year. The presenter summed up by saying: "I like your product, the price is right, just make sure it doesn't break."

There was some murmuring in the audience followed by some corporate hand waving quoting reliability figures, but the message indisputably expresses the down-to-earth sentiments of most customers of any technical product: our expectations are that it will work without failure, whenever called upon to do so, until we ourselves find the product obsolete and outdated and discard it for these reasons.

Of course, life is not simple, and no professional will make claims of product reliability being 100 percent. But how close can we get? Well, this is what all our quality and reliability efforts revolve around.

As we see, the reliability, and the price we are ready to pay for it, are intimately connected topics. As Rzhanitsyn (1978) suggests, one ought to allocate an optimum reliability, corresponding to price minimization process. Examples of detailed calculations of this kind are given by Ang and De Leon (1997), for example, and other authors.

In the following section, we will discuss an alternative, non-probabilistic, method to deal with uncertainty in the same problem.

6.3 REMARKS ON CONVEX MODELING OF UNCERTAINTY

The number of linear problems has been considered under set-theoretical, convex modeling of uncertainty in structures, and in monograph (Ben-Haim & Elishakoff, 1990). Particularly, impact failure of bars (Ben-Haim & Elishakoff, 1989) and shells (Elishakoff & Ben-Haim, 1990) was studied in detail, as was the response of a vehicle in uneven terrain (Ben-Haim & Elishakoff, 1989). By contrast, the only nonlinear problem considered in applied mechanics literature within the set-theoretical, convex modeling is buckling of shells with uncertain initial imperfections (Ben-Haim & Elishakoff, 1989). The latter paper studied the first- and second-order approximation for the nonlinear function, since the exact solution was unavailable. The present section contrasts the first- and second-order approximation discussed in detail by Ben-Haim and Elishakoff (1990). Here, we go beyond the latter monograph, presenting, apparently for the first time, the exact analysis for the model structure of the bar with two eccentricities.

Consider again an elastic bar under an axial compressive force, applied with eccentricities e_1 and e_2. The maximum bending moment M_z^* is given by Equation (6.8).

With e_1 and e_2 specified, the maximum value of the moment can be directly evaluated from Equation (6.8). Assume now that the initial eccentricities are uncertain. In contrast to the previous section, we do not propose to model this uncertainty as randomness, under a probabilistic approach, but use an alternative, set-theoretical description, usually called "convexity modeling." The nominal values of the eccentricities are e_1^0 and e_2^0, respectively, and the deviations from these nominal values are denoted ζ_1 and ζ_2. We assume that these deviations vary within the ellipsoidal set:

$$Z(\alpha, \omega_1, \omega_2) = \left\{ (\zeta_1, \zeta_2) : \left(\frac{\zeta_1}{\omega_1}\right)^2 + \left(\frac{\zeta_2}{\omega_2}\right)^2 \leq \alpha^2 \right\} \tag{6.23}$$

where ω_1 and ω_2 are semi-axes of the ellipsoid, and α is its size parameter. We are interested in finding the maximum $\mu(\alpha,\omega_1, \omega_2)$, with respect to the uncertainty in the eccentricity, of the maximum bending moment

$$\mu(\alpha, \omega_1, \omega_2) = \max M_z^*\left(e_1^0 + \zeta_1,\ e_2^0 + \zeta_2\right); \quad \{\zeta_1, \zeta_2 \in Z(\alpha, \omega_1, \omega_2)\} \tag{6.24}$$

where $\mu(\alpha,\omega_1, \omega_2)$ is the bending moment of the weakest bar in the ensemble Z. The maximum bending moment for uncertain eccentricities ζ_1 and ζ_2 to the first-order in ζ_1 and ζ_2 is

$$M_z^*(e_1^0 + \zeta_1, e_2^0 + \zeta_2) = M_z^*(e_1^0,\ e_2^0) + \left.\frac{\partial M_z^*}{\partial e_1}\right|_{e_i = e_i^0} \zeta_1 + \left.\frac{\partial M_z^*}{\partial e_2}\right|_{e_i = e_i^0} \zeta_2 \tag{6.25}$$

We will evaluate the maximum bending moment as ζ_1 and ζ_2 vary in an ellipsoidal set $Z(\alpha, \omega_{1,} \omega_2)$. For convenience, we define the vector γ as follows:

$$\gamma^T = \left\{ \left.\frac{\partial M_z^*}{\partial e_1}\right|_{e_i = e_i^0},\ \left.\frac{\partial M_z^*}{\partial e_2}\right|_{e_i = e_i^0} \right\} \tag{6.26}$$

where the superscript T denotes matrix transposition. Equation (6.24), in view of Equations (6.25) and (6.26) becomes:

$$\mu(\alpha, \omega_1, \omega_2) = \max_{\zeta_1, \zeta_2 \in Z(\alpha, \omega_1, \omega_2)} \left[M_z^*(e_1^0, e_2^0) + \varphi^T \zeta \right], \tag{6.27}$$

where

$$\zeta^T = (\zeta_1, \zeta_2). \tag{6.28}$$

Define Ω as 2×2 diagonal matrix

$$\Omega = \begin{bmatrix} \dfrac{1}{\omega_1^2} & 0 \\ 0 & \dfrac{1}{\omega_2^2} \end{bmatrix} \tag{6.29}$$

Then Equation (6.23) can be rewritten as

$$A(\alpha, \Omega) = \{\zeta : \zeta^T \Omega \zeta \leq \alpha^2\} \tag{6.30}$$

Equation (6.27) calls for finding the maximum of the linear functional $\gamma^T\zeta$ on the convex set $Z(\alpha, \omega_1, \omega_2)$. According to the well-known theorem (see e.g., Leunberger 1984; Arora, 1989) a linear functional, considered on the convex set Z, assumes the maximum on the set of extreme points of Z. The latter is the collection of vectors $\sigma = (\zeta_1, \zeta_2)$ in the following set:

$$C(\alpha, \Omega) = \{\sigma : \sigma^T \Omega \sigma = \alpha^2\} \tag{6.31}$$

Thus, the maximum bending moment becomes

$$\mu(\alpha, \Omega) = \max_{\sigma \in C(\alpha, \Omega)} [M_Z^*(e_1^0, e_2^0) + \gamma^T \sigma] \tag{6.32}$$

To solve the problem, we use the method of Lagrange multipliers. For details of derivation, the reader should consult the paper by Elishakoff, Gana-Shvili, and Givoli (1991). Probabilistic analysis of the identical problem is given by Elishakoff and Nordstrand (1991).

For the maximum bending moment, we arrive at the following equation

$$\mu(\alpha, \omega_1, \omega_2) = M_Z^*(e_1^0, e_2^0) + \gamma^T \sigma_1 = M_Z^*(e_1^0, e_2^0) + \alpha \sqrt{\gamma^T \Omega^{-1} \gamma} \tag{6.33}$$

In an analogous manner we arrive at the following equation for the minimum bending moment

$$\mu_{\min}(\alpha, \omega_1, \omega_2) = M_Z^*(e_1^0, e_2^0) + \gamma^T \sigma_2 = M_Z^*(e_1^0, e_2^0) - \alpha \sqrt{\gamma^T \Omega^{-1} \gamma} \tag{6.34}$$

For the problem under consideration elements of vector φ can be found analytically:

$$\frac{\partial M_Z^*}{\partial e_1} = \frac{P\beta_1}{\sqrt{\beta_3}\ \sin kL} \tag{6.35}$$

$$\frac{\partial M_z^*}{\partial e_2} = \frac{P\beta_2}{\sqrt{\beta_3}\sin kL} \tag{6.36}$$

where

$$\beta_1 = e_1 - e_2 \cos kL$$
$$\beta_2 = e_2 - e_1 \cos kL \tag{6.37}$$
$$\beta_3 = e_1^2 + e_2^2 - 2e_1 e_2 \cos kL$$

Hence the maximum bending moment reads

$$M_{\max}(\alpha, \omega_1, \omega_2) = M_z^*(e_1^0, e_2^0) \tag{6.38}$$
$$+\alpha\sqrt{\left[\omega_1 \frac{\partial M_z(e_1,e_2)}{\partial e_1}\right]^2_{e_i=e_i^0} + \left[\omega_2 \frac{\partial M_z(e_1,e_2)}{\partial e_2}\right]^2_{e_i=e_i^0}}$$

whereas the minimum bending moment is

$$M_{\max}(\alpha, \omega_1, \omega_2) = M_z^*(e_1^0, e_2^0) \tag{6.39}$$
$$-\alpha\sqrt{\left[\omega_1 \frac{\partial M_z(e_1,e_2)}{\partial e_1}\right]^2_{e_j=e_j^0} + \left[\omega_2 \frac{\partial M_z(e_1,e_2)}{\partial e_2}\right]^2_{e_j=e_j^0}}$$

The detailed second-order analysis can be found in the paper by Elishakoff, Gana-Shvili and Givoli (1991).

We will show below that the maximum bending moment is a convex function of its arguments ζ_1 and ζ_2. Indeed, according to a well-known theorem (Leunberger, 1984; Arora, 1989), a function of n variables defined on a convex set S is convex if and only if its Hessian matrix is positive semi-definite at all points in S. In our case the elements of the Hessian matrix γ_{ij} are

$$\gamma_{11} = \frac{e_2^2(1+\cos kL)^2}{\left[e_1^2 + e_2^2 - 2e_1 e_2 \cos kL\right]^{3/2}} \tag{6.40}$$

$$\gamma_{12} = -(e_1^2 + e_2^2 - 2e_1 e_2 \cos kL)^{-1/2}\left[\cos kL + \frac{(e_1 - e_2\cos kL)(e_2 - e_1 \cos kL)}{e_1^2 + e_2^2 - 2e_1 e_2 \cos kL}\right] \tag{6.41}$$

$$\gamma_{22} = \frac{e_1^2 (1+\cos kL)^2}{\left[e_1^2 + e_2^2 - 2e_1 e_2 \cos kL\right]^{3/2}} \tag{6.42}$$

Direct calculation yields

$$\det[\Gamma] = \gamma_{11}\gamma_{22} - \gamma_{12}^2 \equiv 0 \tag{6.43}$$

Also

$$\gamma_{11} > 0,\ \gamma_{22} > 0. \tag{6.44}$$

Equations (6.43) and (6.44) imply that the function M_Z^* is convex. Therefore, we can apply the following theorem (Leunberger, 1984; Arora, 1989); "Let f be a convex function defined on a bounded closed convex set S. If f has a maximum over S, this maximum is achieved at an extreme point of S."

Such being the case, we deduct that the maximum moment is achieved on the ellipse

$$\left(\frac{\zeta_1}{\omega_1}\right)^2 + \left(\frac{\zeta_2}{\omega_2}\right)^2 = \alpha^2 \tag{6.45}$$

We express ζ_2 from the latter equation as

$$\zeta_2 = \pm\, \omega_2 \sqrt{\alpha^2 - \frac{\zeta_1^2}{\omega_1^2}} \tag{6.46}$$

and substitute in Equation (6.24) to yield

$$M_Z^*(e_1^0 + \zeta_1, e_2^0 + \zeta_2) = M_Z^*\left(e_1^0 + \zeta_1, e_2^0 \pm \omega_2 \sqrt{\alpha^2 - \zeta_1^2 / \omega_2^2}\right) \tag{6.47}$$

Now we seek the maximum of M_Z^* with respect to ζ_1 alone. For the maximum moment we demand that

$$\frac{\partial M_Z^*}{\partial \zeta_1} = 0 \tag{6.48}$$

This equation defines the ζ_1^* at which M_Z^* assumes maximum; then Equation (6.43) determines the value of ζ_2^*. Substituting ζ_1^* and ζ_2^* into Equation (6.24), we obtain the maximum value of the bending moment.

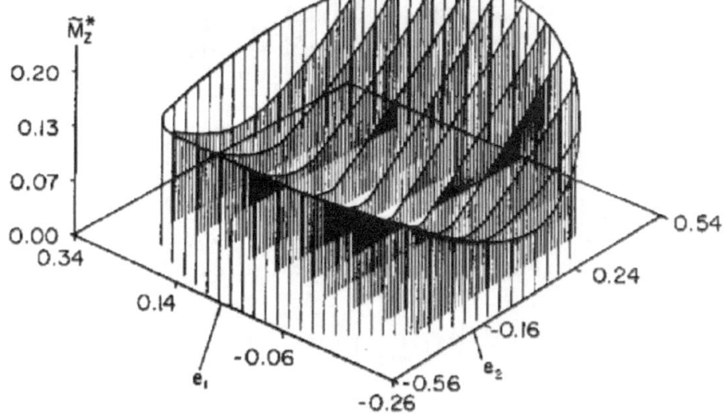

FIGURE 6.7 Variation of the bending moment over the uncertainty ellipse

Figure 6.7 shows the variation of the bending moment over the region:

$$\left(\frac{\zeta_1}{\omega_1}\right)^2 + \left(\frac{\zeta_2}{\omega_2}\right)^2 \leq \alpha^2 \tag{6.49}$$

for $\omega_1 = 0.3$, $\omega_2 = 0.5$, at the non-dimensional load level

$$v = \frac{P}{P_E}, \; P_E = \frac{\pi^2 EI}{L^2} \tag{6.50}$$

equal n = 0.3. Figure 6.8 is associated with $\omega_1 = 0.3$, $\omega_2 = 0.6$ and n = 0.3, whereas Figure 6.9 illustrates the variation of M_Z^* for the values $\omega_1 = 0.3$, $\omega_2 = 0.6$ and n = 0.5. As we see in all these three instances the maximum value is achieved at the boundary point of the ellipse.

The variation of the non-dimensional maximal moment

$$\overline{M}_z^* = \frac{M_Z^*}{P_e c} \tag{6.51}$$

versus the ellipse's size α, where *c* is the radius of inertia of the bar's cross-section is given in Figure 6.10. Moreover, $\omega_1 = 0.001$, $\omega_2 = 0.02$, $e_1^0 = 0.04$, $e_2^0 = 0.04$. The non-dimensional load is fixed at n = 0.3.

Broken curves marked 1 are associated with the first-order analysis, curves marked 2 with the second-order analysis and curves marked 3 with the exact results. The maximum moment increases with the size α of the uncertainty ellipse. For moderate values of α the agreement between the first-order, second-order and the exact analysis is excellent. It turns out that with the increase of the

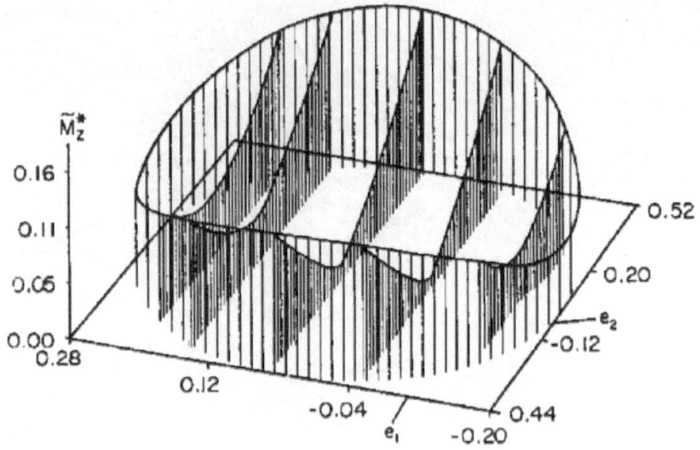

FIGURE 6.8 Moment uncertainty as a function of uncertainty in eccentricities

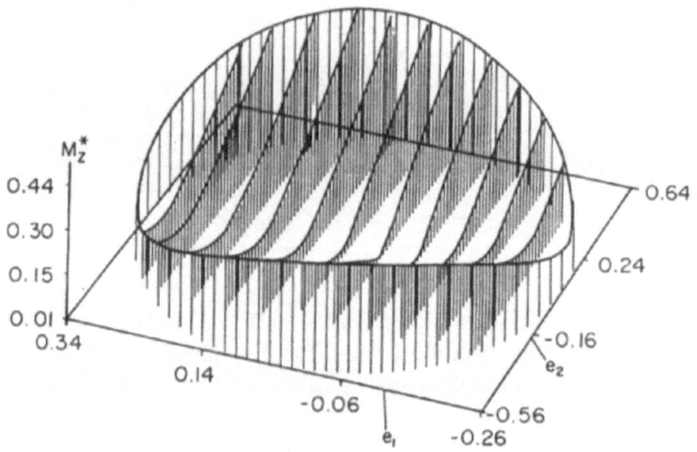

FIGURE 6.9 Moment uncertainty as a function of uncertainty in eccentricities

non-dimensional applied load, the percentage-wise disagreement between the first-order and the second-order analyses decreases considerably. On the other hand, for the larger non-dimensional applied loads the difference between the low-order approximations and the exact analysis becomes wider.

We conclude, therefore, that the lower-order approximations yield acceptable results for small uncertainties and smaller loads. This modifies the conclusion, based on the comparison of the first-order and the second-order analyses, drawn in by Ben-Haim and Elishakoff (1990) that the first-order approximation was acceptable for small uncertainties and greater loads. It is remarkable that the similarity

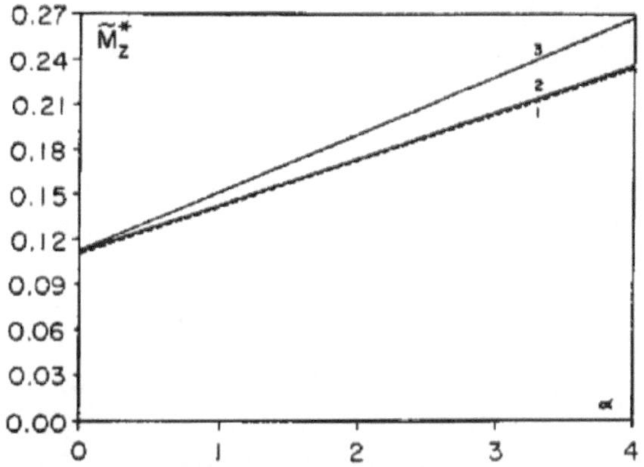

FIGURE 6.10 Comparison of first-order and second-order solutions with exact solution

of the first-order and second-order approximations that occurs for the elastic bar does not generally suggest that these approximations are in good agreement with the exact solution.

6.4 "WORST-CASE" PROBABILISTIC SAFETY FACTOR

The following question arises: It is possible to perform worst-case analysis within the probabilistic context? At first glance this question may appear to be not valid. Indeed, if the random loads involved may take values from $-\infty$ to $+\infty$, naturally there is no "worst" value of the load. Some engineers suggest, therefore, to utilize some characteristic "worst" case analysis.

However, in practical applications, precisely, such an approach is often adopted. Bolotin (1968) notes:

> In engineering codes as a design load some "maximum" (i.e. corresponding to certain small probability of realization) value is taken, whereas as a design resistance of the material some lower value, determined from technical conditions is utilized. Choice of these values, as well as of safety factor, is in some degree arbitrary: Various values of safety factor could correspond to some value of reliability.

Consider the following example treated by Maymon (1998). We will study the probabilistic behavior of an elastic bar of the rectangular cross-section with width B and depth H. The bar is under the tensile force F. The quantities B, H, and F are all normally distributed with

$$\begin{aligned} \sigma_B &= 0.01\,E(B) \\ \sigma_H &= 0.01\,E(H) \\ \sigma_F &= 0.01\,E(F) \end{aligned} \tag{6.52}$$

The normal stress in the bar

$$\Sigma = F / BH \tag{6.53}$$

is, naturally, not distributed normally. Yet, due to small standard deviations involved, it can be approximately treated as a normally distributed random variable. If $E(\Sigma) = 2083$ kg/cm^2 and $\sigma_\Sigma = 104$ kg/cm^2, then we can conditionally agree to define the maximum tensile stress as the mean value of the stress *plus* three times the standard deviation of it:

$$\sigma_{max} = 2083 + 3 \times 104 = 2395 \text{ kg/cm}^2 \tag{6.54}$$

The failure of the bar is identified with the exceedance by the tensile stress the yield stress. The latter, Σ_y, is also assumed to be normally distributed with

$$\begin{aligned} E(\Sigma_y) &= 2500 \text{ kg/cm}^2 \\ \sigma_{\Sigma_y} &= 75 \text{ kg/cm}^2 \end{aligned} \tag{6.55}$$

We introduce the conditionally "minimum" yield stress as the mean value *minus* three times its standard deviation:

$$\sigma_{min} = 2500 - 3 \times 75 = 2275 \text{ kg/cm}^2 \tag{6.56}$$

According to Maymon (1998), "three types of designers who would treat this problem differently are identified."

(a) The "nominal designer" evaluates the safety factor of the bar from nominal, mean values

$$n_{nominal} = E(\Sigma_y) / E(\Sigma) = 2500/2083 = 1.2 \tag{6.57}$$

which coincides with the central safety factor. If this value is not less than the required factor of safety specified for the specific project, the design is accepted. However, the design neglects the scatter expected during the production phase in variables B, H, F, and Σ_y, which constitute random variables.

(b) The second type of the designer is identified by Maymon (1998) as the "3σ worst case designer." Such a designer calculates the safety factor as follows

$$n_{3\sigma} = \sigma_{y,min} / \sigma_{max} = 2275 / 2395 = 0.95 \tag{6.58}$$

which is not acceptable, since it is less than unity. One can make a correction in design. Increasing of the cross-section by 26% obtainable through forming the ratio 1.2/0.95 = 1.26 increases the weight by 26% and, consequently, the cost of the product. Another possibility is to decrease the allowed tolerances in B and H. One can show that in order to obtain the unity safety factor, which is less than the required value of 1.2, the tolerances on the cross-section should be decreased from 1% to 0.04%. Even this change, however, may contribute to the tremendous increase in the product cost.

(c) The third type of designers is utilizing the reliability concept directly. Assume that the required value of reliability is 99.9%, then the "probabilistic designer" (in Maymon's terminology) uses the probabilistic analysis. Probabilistic analysis demonstrates that

$$Prob\left(\Sigma < \Sigma_y\right) = 99.943\% \tag{6.59}$$

Since this value is higher than the required 99.9% value, the designer is allowed to somewhat decrease the cross-section or slightly increase the distance on the cross-section area.

Let us prove the validity of Equation (6.59) (Maymon, 2002b). The failure function g delimiting the safe and unsafe regions of operation reads

$$g\left(\Sigma, \Sigma_y\right) = \Sigma_y - \Sigma = 0 \tag{6.60}$$

We transform the discussion into the space of standardized variables

$$\begin{aligned} Z_1 &= \frac{\Sigma_y - E\left(\Sigma_y\right)}{\sigma_{\Sigma_y}} \\ Z_2 &= \frac{\Sigma - E(\Sigma)}{\sigma_\Sigma} \end{aligned} \tag{6.61}$$

Substituting Equation(6.61) into Equation (6.60) we get

$$G\left(z_1, z_2\right) = \sigma_{\Sigma_y} Z_1 - \sigma_\Sigma Z_2 + E\left(\Sigma_y\right) - E(\Sigma) = 0 \tag{6.62}$$

Hereinafter, we use the Lagrange multiplier method. Since the minimum distance from the origin to the failure boundary given in Equation (6.60) is given by equation

$$d = \sqrt{z_1^2 + z_2^2} \tag{6.63}$$

we construct the following Lagrangean:

$$D = Z_1^2 + Z_2^2 - \lambda\left[\sigma_{\Sigma_y} Z_1 - \sigma_\Sigma Z_2 + E\left(\Sigma_y\right) - E(\Sigma)\right] \tag{6.64}$$

where λ is the Lagrange multiplier. Equation (6.62) in addition to the following two equations

$$\frac{\partial D}{\partial Z_1} = 2Z_1 - \lambda\sigma_{\Sigma_y} = 0 \tag{6.65}$$

$$\frac{\partial D}{\partial Z_2} = 2Z_2 + \lambda\sigma_\Sigma = 0 \tag{6.66}$$

provides three equations for three values: Z_1, Z_2 and the value of λ. We get

$$\begin{aligned} Z_1 &= -1.9022566 \\ Z_2 &= 2.63779575 \\ \lambda_3 &= -0.05072684 \end{aligned} \tag{6.67}$$

The reliability index becomes

$$\beta = \sqrt{Z_1^2 + Z_2^2} = 3.25216 \tag{6.68}$$

which corresponds to probability of failure 0.0572719 by using

$$P_f = \Phi(-\beta) \tag{6.69}$$

or a reliability of 99.943% as it was required to demonstrate.

Maymon (1998) concludes:

> Although this example is a simplification of real-life application, it shows that contradictory measures can be taken when using each approach. It also indicates that the probabilistic approach may be more realistic and may result in less conservative design. This approach is valid if project requirements are formulated by probabilistic methods. Although this is not the situation at present, it is believed that future requirements will contain more and more of this approach.

It should be stressed that it is not necessary the worst-case design is to be associated with the mean plus or minus three standard deviations. Consider the following example due to Hart (1982).

A truss member of area a is subjected to an axial load P. Since P is given as a normal random variable, so is the resulting stress, denoted as Σ: $\sigma = P/a$. Let Σ_y, can be also a normal random variable, independent of P. The probabilistic characteristics are

$$\begin{aligned} E(P) &= 200\ kips, \quad \sigma_P = 40\ kips \\ E(\Sigma_y) &= 47.9\ ksi, \quad \sigma_{\Sigma y} = 3.3\ ksi \end{aligned} \tag{6.70}$$

The member is designed in such a manner the central safety factor constitutes 1.2. We need to calculate the required truss member area. The central safety factor is

$$n = 1.2 = \frac{E(\Sigma_y)}{E(P)/a} \tag{6.71}$$

From here we find

$$a = 1.2\frac{E(P)}{E(\Sigma_y)} = \frac{1.2 \times 200}{47.9} = 5.01\ in^2 \tag{6.72}$$

Deterministic design load is defined as follows

$$P_{design} = E(P) + \alpha_P \sigma_P \tag{6.73}$$

when α_P is a constant such that the probability that the load is greater than P_{design} is 0.10. We observe that the probability that the load does not exceed that value P_{design} equals 0.9. Hence,

$$Prob(P \le P_{design}) = 0.9 = \int_{-\infty}^{P_{design}} \frac{1}{40\sqrt{2\pi}} \exp\left[-\frac{1}{2}\left(\frac{t-200}{40}\right)^2\right] dt \tag{6.74}$$

Standardized normal-distribution table shows that

$$P_{design} = 251.3\ kips \tag{6.75}$$

Hence,

$$P_{design} = E(P) + \alpha_P \sigma_P = 200 + \alpha_P \times 40 = 251.3 \tag{6.76}$$

Thus,

$$\alpha_P = 1.28 \tag{6.77}$$

We represent P_{design} as follows:

$$P_{design} = \psi_{Load} E(P) \tag{6.78}$$

where the coefficient ψ_{Load} is called load factor. Substitution yields

$$P_{design} = 251.3 = \psi_{Load} E(P) = \psi_L \times 200 \tag{6.79}$$

resulting in

$$\psi_{Load} = 1.26 \tag{6.80}$$

We also define the deterministic design yield stress level to be

$$\sigma_{y,design} = E(\Sigma_y) - \alpha_{\Sigma_y} \sigma_{\Sigma_y} \tag{6.81}$$

where the coefficient α_{Σ_y} is defined so that the probability that the yield stress is less than the designed value, equals 0.1. This leads to

$$0.1 = Prob(\Sigma_y \le \sigma_{y,design}) = \int_{-\infty}^{\sigma_{y,design}} \frac{1}{3.3\sqrt{2\pi}} \exp\left[-\left(\frac{t-47.9}{3.3}\right)^2\right] dt \tag{6.82}$$

Using again the standardized normal distribution table we find

$$\sigma_{y,design} = 43.7\ ksi \tag{6.83}$$

Hence,

$$\sigma_{y,design} = E(\Sigma_y) - \alpha_{\Sigma_y} \sigma_{\Sigma_y} = 47.9 - \alpha_{\Sigma_y} \times 3.3 = 43.7 \tag{6.84}$$

leading to

$$\alpha_{\Sigma_y} = 1.28 \tag{6.85}$$

This allows one to introduce the capacity reduction factor $\psi_{capacity}$, if we define

$$\sigma_{y,design} = \psi_{capacity} E(\Sigma_y) \tag{6.86}$$

leading to, numerically,

$$43.7 = \psi_{capacit} \times 47.9 \tag{6.87}$$

Hence,

$$\psi_{capacity} = 0.91 \tag{6.88}$$

Now we can determine the required truss member area such that

$$\sigma_{design} = \frac{P_{design}}{a} \tag{6.89}$$

Thus,

$$a = \frac{P_{design}}{\sigma_{design}} = \frac{251.3}{43.7} = 5.75\ in^2 \tag{6.90}$$

Now we can answer the question: "What is the probability of failure if the latter is defined to be a load-induced stress equal to or greater than a material yield stress when the area of the truss member is determined" (Hart, 2002). To answer this question, we introduce the failure boundary function

$$g = \Sigma_y - \frac{P}{a} \tag{6.91}$$

Now,

$$\begin{aligned} E(g) &= E(\Sigma_y) - \frac{E(P)}{a} \\ \sigma_g^2 &= \sigma_{\Sigma_y}^2 + \frac{\sigma_P^2}{a^2} \end{aligned} \tag{6.92}$$

The probability of failure is, therefore,

$$P_f = \int_{-\infty}^{\sigma} \frac{1}{\sigma_g \sqrt{2\pi}} \exp\left[-\frac{1}{2} \left(\frac{t - E(g)}{\sigma_g} \right)^2 \right] dg \tag{6.93}$$

Calculation yields

$$P_f = 0.178 \tag{6.94}$$

Likewise, if

$$a = 5.75\ in^2, E(g) = 13.12, \sigma_g = 7.70 \tag{6.95}$$

then

$$P_f = 0.044 \tag{6.96}$$

Hart (2002) summarizes the evaluation of this example as follows:

> Note that the two alternative deterministic design approaches result in two different member areas and hence two different probabilities of failure. The second approach directly incorporates the uncertainty in the loading and strength of the system.

6.5 WHICH CONCEPT IS MORE FEASIBLE: NON-PROBABILISTIC RELIABILITY OR CONVEX SAFETY FACTOR?

In recent decades the non-probabilistic reliability concept was advocated by Ben-Haim (1994). He writes:

> Reliability has a plain lexical meaning, which the engineers have modified and absorbed into their technical jargon. Lexically, that which is "reliable" can be depended upon confidently. Applying this to machines or system, they are "reliable" (still avoiding technical jargon) if one is confident that they will perform their specified tasks as intended. In current technical jargon, a system is reliable if the probability of failure is acceptably low. This is legitimate extension of the lexical meaning, since "failure" would imply behavior beyond the domain of specified tasks. The particular innovation which makes the development of modern engineering reliability is the insight that probability – a mathematical theory – can be utilized to quantify the qualitative lexical concept of reliability.

Ben-Haim (1994) further advocates:

> We do not detract from the importance of the probabilistic concept of reliability by suggesting that probability is not the only starting point for quantifying the intuitive idea of reliability. Probabilistic reliability emphasized the *probability* of acceptable behavior. Non-probabilistic reliability, as developed here [Ben-Haim 1994], stresses the *range* of acceptable behavior. Probabilistically, a system is reliable if the probability of unacceptable behavior is sufficiently low. In the non-probabilistic formulation of

reliability, a system is reliable if the range of performance fluctuations is acceptably small.

In another study (1997) the following statement is made:

> The current standard theory of reliability is based on probability: The reliability of a system is measured by the probability of non-failure ... In this paper we will describe a different formulation. We measure the reliability of a system by the amount of uncertainty consistent with no-failure. A reliable system will perform satisfactorily in the presence of great uncertainty. Such a system is robust with respect to uncertainty, and hence the name robust reliability.

In the discussion of the above 1994 paper one of the present writers [1.E] (1995) noted:

> It appears ... that the non-probabilistic concept of reliability is not necessary .... Engineers are accustomed to highly reliable structures, with reliabilities of order $1–10^{-7}$, which has a frequency interpretation if the ensemble of produced structures is sufficiently large. A natural question arises: will non-probabilistic reliability take on comparable values? ... Indeed, there exists a universally accepted probabilistic definition of reliability. The alternative, non-probabilistic definition of reliability is not formulated ... What one may need, however, is the non-probabilistic concept of safety factor.

The question arises: How to introduce the non-probabilistic safety factor? To reply to this inquiry, consider a system subjected to a combination of loads which belongs to some convex set. Assume that the mathematical model is convex too. By utilizing convex analysis we determine the interval of variation of the stress Σ:

$$\Sigma = \left[\underline{\sigma}, \overline{\sigma}\right] \tag{6.97}$$

where $\underline{\sigma}$ is the minimum stress that the system may experience, when loads vary in the convex set, whereas $\overline{\sigma}$ is the maximum stress that the system may assume. Let the yield stress Σ_y also be an interval uncertain variable

$$\Sigma_y = \left[\underline{\sigma}_y, \overline{\sigma}_y\right] \tag{6.98}$$

Then the safety factor n can be defined in two different ways. One of the possible definitions is as a ratio of two interval variables

$$N = \Sigma_y / \Sigma \tag{6.99}$$

Thus, the safety factor turns out to be an interval variable

$$N = \left[\underline{n}, \overline{n}\right] \tag{6.100}$$

where $\underline{n}$ is the minimum value N may assumes while $\bar{n}$ is the maximum value. Naturally,

$$\underline{n} = \underline{\sigma}_y / \bar{\sigma}, \qquad \bar{n} = \bar{\sigma}_y / \underline{\sigma} \tag{6.101}$$

The definition was proposed by Elishakoff (1995). Another definition is an analogue probabilistic central safety factor, and could be defined as follows

$$n = \frac{\left(\bar{\sigma}_y - \underline{\sigma}_y\right)/2}{\left(\bar{\sigma} - \underline{\sigma}\right)/2} \tag{6.102}$$

where the numerator represents the mid-value of the yield stress interval, while the denominator constitutes a mid-value of the stress interval.

Still, a nagging question may present: Can one define the non-probabilistic reliability? It is noted that the papers (Ben-Haim 1994, 1997) did not introduce the formal definition, although advocated for it. In the paper by Elishakoff (1995) the following possible definition was suggested:

$$R = 1 - 1/n \tag{6.103}$$

for the required safety factor always exceeds unity.

The definition implies that

$$P_f n = 1 \tag{6.104}$$

where $P_f = 1 - R$ is a probability of failure, or unreliability. In other words, one could define a non-probabilistic unreliability as the reciprocal of the non-probabilistic safety factor. Still, such a definition may still not appeal to the practicing engineers: In order to achieve the level of non-probabilistic reliability of, say 0.9, one needs a safety factor of 10! It appears that this observation, as well as the drawbacks of other plausible definitions, is the main reason why the engineering community may not adopt the non-probabilistic reliability.

One may ask a natural question: Why propose a possible definition of the non-probabilistic reliability, if one is skeptical of its very usefulness? In this a "linguistic quibble," as it may appear of the first glance?

It is felt that this is not so; we followed here ancient sages who would strengthen the other point of you, by a more valid argument, but then still would disagree. With these observations, one must strongly disagree with the radical opinion of Good (1996) maintaining that the Ben-Haim's concept of non-probabilistic reliability is an "oxymoron."

In another paper, Good (1995) maintains, as the telling title of his study suggests, "reliability always depends on probability of course." Whereas, in general, this statement appears to be valid, still, the dependence of the reliability upon probability does not exclude its dependence on other concepts. Elishakoff and Colombi (1993) combined probabilistic analysis with convex modeling to obtain the upper and lower bounds of the displacement variance of a structure. Elishakoff and Li (1999) combined probabilistic and convex modeling techniques to derive upper and lower bounds of the probability of failure. What Good (1995, 1996) is against, then, is the exclusively non-probabilistic treatment of the reliability by Ben-Haim (1994, 1997).

For six different levels of treatment of uncertainties in risk analysis one should consult the insightful article by Paté-Cornell (1996).

The "excursion" to non-probabilistic reliability is still useful, for it at least elucidates various possible approaches to the problem of reliability. As Berg (1961) maintained, the "reliability is problem number one." If so, various possible hybrid approaches to it may be pursued, even if the professional consensus presently accepts only a single one.

More recently, Ben-Haim (2001) discusses the information-gap uncertainty:

> Info-gap models quantify uncertainty as a size of the gap between what is known and what could be known.

Here, immediately, numerous questions arise on why the information that "could be known" is not presently "known"? Is this due to negligence of the particular designer or of the entire profession which did not think it was necessary to "know" what "could be known," or the price involved with obtaining the information that "could be known" is too heavy presently, or because this needed information is not released by appropriate firms and companies? It appears that such an analysis, if defined as above, should include an answer to the posed question. Ben-Haim (2001) states:

> Info-gap models of uncertainty are particularly useful when data on the uncertainties are quite limited.

One needs to know how "limited" the information should be to utilize info-gap models. The following pertinent questions arise: Why for the data that is available and considered as appropriate for the info-gap models, one cannot use the probabilistic modeling? Why not to utilize in such circumstances the fuzzy sets-based approaches? Is there a difference between the convex modeling of uncertainty (Ben-Haim and Elishakoff 1990) and the information-gap uncertainty? Is the "information-gap uncertainty" a new terminology or a new methodology? As we see there are plenty of questions demanding further research and answers.

Media and engineering professions simultaneously talk about the "age of information," "information highway," and "information overloading." Why is it that in the engineering profession we are still lacking information on probabilistic distributions of loads or properties of the materials? The former Vice President of USA. Al Gore (1995) wrote:

> we will connect and provide access to the National Information Infrastructure for every classroom, every library, and every hospital and clinic in the entire United States of America.

Engineers are dreaming about the time when the international data banks of the material properties will be available for those who want to check validity of probabilistic, fuzzy sets-based or convex analyses. One should mention that some steps are already undertaken in this direction (see, e.g., Arbocz 1982; Arbocz & Abramovich 1979; Singer et al, 1978; Scherrer & Schuëller, 1988; Cooke et al., 1993; Cooke, 1996).

6.6 CONCLUDING COMMENTS ON HOW TO TREAT UNCERTAINTY IN A GIVEN SITUATION

This chapter reviews some pertinent questions associated with uncertainty modeling in analysis of structures. It gives a critical appraisal of the probabilistic method and describes a new, non-probabilistic philosophy. The former is valid when plentiful information (probability densities) is available on the uncertain quantities involved. The latter convex modeling is appropriate when the existing data is scarce, and no valid probabilistic models can be constructed. Both of these approaches deal with different facets of uncertainty treatment. Theory of probability and random processes is not the only way to deal with uncertainty. Indeed, as Freudenthal (1956) notes:

> ignorance of the cause of variation does not make such variation random.

These two methods appear to successfully complement each other, to make useful judgments based on reliably available experimental information. The new safety factor based on convex modeling is introduced.

It is apropos to quote from Benjamin and Cornell (1970):

> How engineer chooses to treat uncertainty in a given situation depends upon the situation. If the degree of variability is small, *and* if the consequences of any variability will not be significant the engineer may choose to ignore it by simply assuming that the variable will be equal to the best available estimate. This estimate might be the average of the number of past observations. This is typically done, for example, with the elastic constants of materials and the physical dimensions of many objects.
>
> If, on the other hand, uncertainty is significant, the engineer may choose to use a "conservative estimate" of the factor. This has often been done, for example, by setting "specified minimum" strength properties of materials … Many questions arise in the practice of using conservative estimates. For example:
>
> How can engineer maintain consistency in their conservatism from one situation to another? For example, separate professional committees set the specifies minimum bending strength of wood.

It was demonstrated by Elishakoff, Li, and Starnes (2001) that in some situations there may be *no difference* between the probabilistic and convex models of uncertainty. In certain situations, both approaches tend to yield close numerical results. If this intriguing statement is correct the following conclusions can be made:

(a) One can continue doing the uncertainty analysis of one's choice.
(b) The pragmatic approach would suggest that the analysis that leads to less calculations should be preferred. Naturally, the convex modeling does not involve the probability densities. But one can pose valid questions or find the bounding ellipsoids containing the experimental information.

Still, one has to bear in mind the idea propagated by Thoft-Christensen and Baker (1982):

> Upper limits to the individual loads and lower limits to material strength are not easily identified in practice, e.g., building occupancy loads, wind loads, the yield stress of steel, the cube or cylinder stress of concrete.

It appears instructive to conclude the discussion on non-probabilistic approaches on the safety factors by a quote from Bodner (2003):

> I think it is generally understood that the concept of Factor of Safety (FS) in a measure of the difference between the normal operating conditions (or the maximum in some sense) of a system and the minimum conditions for which the system cannot function. That could be due to many factors such as excessive deformations or cessation of load carrying ability due to inadequacies of materials or structure or to extreme loadings. The determination of the appropriate FS is part of the design process and is usually specified in codes and design handbooks. Of course, the recommended FS is different for various applications such as aircraft, buildings, bridges, underwater vehicles, etc. In many cases, the specific FS depends on the design parameters.
>
> A particular meaning of FS suggested by Plasticity investigators is that the FS of a structure is the collapse condition (load) divided by a geometrically similar load that would initiate yield at some point in the structure. This concept has been generalized and has been used in practical applications.
>
> I know there is much interest in formulating FS based on statistical consideration and various criteria for "probability of failure" have been proposed. These may be useful in some cases. From my observations, it seems that failures are usually due to a particular defect or anomaly in the system or loading condition that was not taken into account in the design. In modern times, misuse of a computer program has led to problems.

Works in this direction are numerous and will not be listed here. In this connection, Bushnell (2003) notes:

> One may ask, "If one performs the collapse analysis, does not he in essence determine the exact value of the safety factor?" A quick answer is "No", because a safety factor can cover for a multiple of sins: unknown imperfections, unknown material properties, etc. In principle, if one knows exactly what the structure and its loading and boundary conditions are, one could do an "exact" finite element analysis and compute the collapse load. Then one could do some sort of approximate analysis, say a bifurcation

buckling analysis of the idealized perfect structure with nominal material properties and boundary conditions and loading. A factor of safety could then be computed as the ratio:

$$\frac{\text{collapse load of actual structure}}{\text{bifurcation buckling of nominal system}}$$

Also, one might assign different factors of safety corresponding to differentiate phenomena, such as *fs*(1) for stress, *fs*(2) for buckling, *fs*(3) for vibration, etc. The various factors of safety, *fs*, would depend upon how well known the phenomena is and how sensitive the predictions are to the variations of the input data.

For further details, the reader may consult the paper by Bushnell (1990).

We ought to note the books devoted to convex modeling of uncertainty. These include the books authored or edited by (in chronological order) Ben-Haim (1996), Elishakoff, Lin, and Zhu (1994), Elishakoff, Li, and Starnes (2001), Hlaváěek, Chleboun, and Babuška (2004), Elishakoff (2004), Elishakoff and Ohsaki (2010), Natke and Ben-Haim (1997), Takewaki (2007), Abbas (2012), Takewaki, Abbas, and Fujita (2013), El Hami and Radi (2013), Lyons (2013), Cursi and Sampao (2015),Lorkowski and Kreinovich (2017), Lodwick and Thipwiwatpotjana (2017), Tilahun and Ngnotchouye (2018), Vasile (2020), Takewaki and Kojima (2021).

7 Systems and Components

The whole is more than the sum of its parts.

Aristotle 384 B.C.E.–322 B.C.E., "Metaphysics"

7.1 CONDITIONALITY OF CONCEPT

A system is usually an aggregate of elements joined together for specific operations while each element is the simplest part of the aggregate. The concepts of system and components utilized in the theory of reliability depend on many aspects. The same object can play the role of a system or an element (component) depending on the problem's specifics. It should also be noted that a structural material itself is an aggregate of crystals, granules, fibers, or other particles and shall be considered as a system in material engineering. More detailed distinction of elements is generally not necessary in structural engineering. Instead, elements or components of a system are bars, beams, plates, cables, panels, blocks, and so on. These components are assembled in a structure in the form of frames, hanging systems, trusses, and other classic systems of structural mechanics.

The role of a system can be played by a population of structures intended for performing specific function(s) in ensemble. Buildings, aircraft, ships, vehicles, and so on are typical examples of structural systems. Thus, from a viewpoint of reliability analysis, it is always possible to establish a hierarchy, and obtain all solutions necessary at each hierarchy level. With respect to reliability the important feature of a system is that for any combination of structural members a failure of one or more of them will lead to operational malfunction of the whole system. The notion of system can include a multitude of structures which do not necessarily operate jointly, for example a number of separate structures of the same design. Indeed, a serious accident with one or a number of these products often leads to inspecting all fabricated party of this type. As an example, from the automotive industry, when a defect is found in a detail or a sub-system which had resulted in, or can potentially lead to, a failure or an accident, all cars of the affected model are recalled for repair.

If a failure of an arbitrary component of a structural system leads to failure of the entire system, this indicates that the system consists of elements connected in

DOI: 10.1201/9781003265993-7

series. If the same failure doesn't lead to failure of the whole system, this implies that the failed element had a parallel connection within the system.

7.2 CONNECTION OF COMPONENTS IN SERIES

In a system with elements connected in series (Figure 7.1), failure of an arbitrary member leads to a failure of the whole system. A statically determinate system can present this type of connection. If a member of this system fails, the system is transformed into a mechanism. Reliability of such systems is less than the reliability of any single member, since failure of any single member out of many of them is more probable than the failure of a given element.

Probability of an element to fail can be denoted as $P_{fi}(r \le q) = P_{fi}(q)$, where r is the resistance factor and q is the applied load. Probability of no failure under load q is equal to $[1 - P_i\,(q)]$. For the system under consideration the probability of no failure is:

$$P_s(q) = 1 - P_f(q) = \prod_{i=1}^{n} \left[1 - P_{fi}(q)\right] \tag{7.1}$$

where $P_f(q)$ is the probability that under a load not exceeding q, one of elements fails and the whole system fails too. It was assumed in (7.1) that random resistances of all elements are independent of each other. If the distributions of all elements' resistances are similar and expressed in the load units (all cross-sections are designed to the same stress), then (7.1) can be replaced by:

$$P_s(q) = [1 - P_{fi}(q)]^n \tag{7.2}$$

Probability density function is:

$$p_f(q) = dP_f \,/\, dq = -\frac{d}{dq}\left\{\prod_{i=1}^{n}\left[1 - P_f(q)\right]\right\} = \sum_{i=1}^{n} \frac{P_{fi}}{1 - P_{fi}} \prod_{i=1}^{n} (1 - P_{fi}) \tag{7.3}$$

FIGURE 7.1 Elements in series

Probability density function for resistance of an arbitrary element is $p_i(q) = dP_i/dq$. If stresses in all elements are equal, then probability density function becomes:

$$p_f(q) = n[1 - P_i(q)]^{n-1} p_i(q) \tag{7.4}$$

Consider, for example, a system consisting of 25 components. Strength of each component is a normally distributed random variable and reliability index $\beta = 3$, that is, $P_s(q) = 0.99865$. It follows from equation (7.2) that $P_i(q) = 1 - \sqrt[25]{0.99865} = 0.000054$ From the table of probability integrals one can get $\Phi(\beta_i) = 0.5–0.000054 = 0.49946$, and $\beta_i = 3.87$. As mentioned before, the reliability of a system is less than the reliability of each of its components. The greater the numbers of elements in the system, n, the less is the failure probability (Figure 7.2). Here the probability of no failure, P_n, is a function of n and represents reliability of the chain consisting of independent links with equal probability of failure P_1.

It was assumed that resistances of the elements are uncorrelated random variables. Validity of this assumption was demonstrated in (Raizer, 1995), where it was shown that for a majority of structural systems with a rather high level of reliability ($\beta \geq 3$), the influence of correlation is very weak. It can also be mentioned that the structural members don't always coincide with the elements of the system considered. The truss chord, for example, can be seen as a simple bar intersecting some nodes and can be considered as one element. The cross brace combined from two bars can also be treated as a single element. A model of in series connection can be illustrated by simulating a simply supported beam under bending as a multi-element chain, whose links sustain the bending moments (Rzhanitsyn, 1978]) (see Figure 7.3).

In contrast with a flexible chain under tension, forces in the links are not equal but proportional to the bending moments in the treated beam. The reliability of the ith link is:

$$P_{si} = 1 - P_f(m_i) \tag{7.5}$$

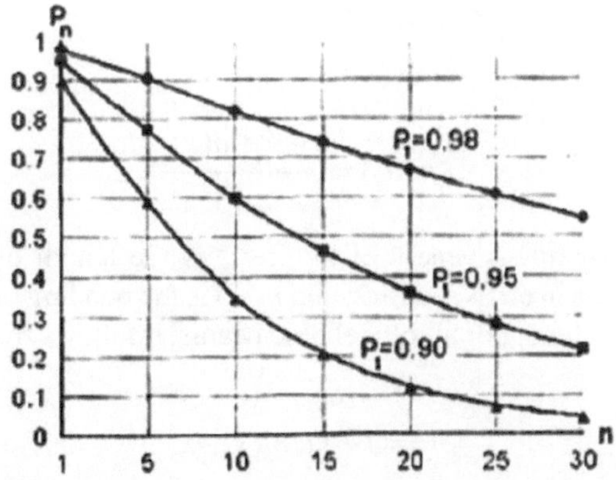

FIGURE 7.2 No failure probability for a system with elements connected in series

FIGURE 7.3 Modeling in series connection by a chain beam under bending

where m_i is moment in the ith link; $P_f(m_i)$ is the probability that the ith link fails when $m \leq m_i$. Reliability of the entire beam for uncorrelated resistances in different links reads:

$$P_s = \prod_{i=1}^{n}[1 - P(m_i)] \tag{7.6}$$

where n is the number of links. Equation (7.6) can be transformed into:

$$\ln P_s = \sum_{i=1}^{n} \ln[1 - P(m_i)] \tag{7.7}$$

If $P(m_i)$ is a small value, then

$$\ln[1 - P_f(m_i)] \approx -P_f(m_i) \tag{7.8}$$

$$\text{and } P_s = \exp\left[\sum_{i=1}^{n} P_f(m_i)\right] \tag{7.9}$$

For a continuous beam with span equal to l:

$$P_s = \exp\left\{-\int_0^1 u[m(x)]dx\right\} \tag{7.10}$$

where

$$u[m(x)] = \frac{1 - P_f[m(x)]}{\lambda}$$

u[m(x)] is the linear intensiveness of failure, λ is the length of beam with fully correlated resistances in cross sections, and *m (x)* is the bending moment curve with regard to the applied load. For highly reliable beams, it follows from Equation (7.6):

$$P_s = 1 - \int_0^1 u[m(x)]dx \tag{7.11}$$

If resistance is normally distributed in cross sections along the beam, then:

$$\left.\begin{aligned} 1-P(m) &= 0.5-\Phi\left[\frac{m-\bar{m}}{s_m}\right] \\ u[m(x)] &= \frac{1}{\lambda}\left\{0.5-\Phi\left[\frac{m-\bar{m}}{s_m}\right]\right\} \end{aligned}\right\} \tag{7.12}$$

The probability of failure is

$$P_f = \frac{1}{2\lambda} - \frac{1}{\lambda}\int_0^1 \Phi\left[\frac{m-\bar{m}}{s_m}\right]dx \tag{7.13}$$

It can be mentioned that for statically indeterminate systems the failure of even one of its members is a signal of danger for the whole system. As a result, reliability can be estimated by treating such systems as a series of connected elements.

7.3 PARALLEL CONNECTION OF BRITTLE COMPONENTS

In contrast with connection in series, the reliability of a system with a parallel connection of elements will be greater than the reliability of any element. If elements are brittle and the limit of resistance in one element is reached, the element would fail due to brittle rupture and stresses in other elements would increase to pick up the load applied to the failed element. This situation can result in the consecutive rupture of all elements. Let's consider two brittle bars connected in parallel as shown in Figure 7.4 on the left side.

The probability of a single bar to fail under force q is $P_f(q)$. The system's reliability should be calculated using the concept of conditional probability:

$$P_s(q) = 1-P_f(q) = [1-P_f(0.5q)][1-P_f(0.5q)] + 2P_f(0.5q)[1-P_f(q)] \tag{7.14}$$

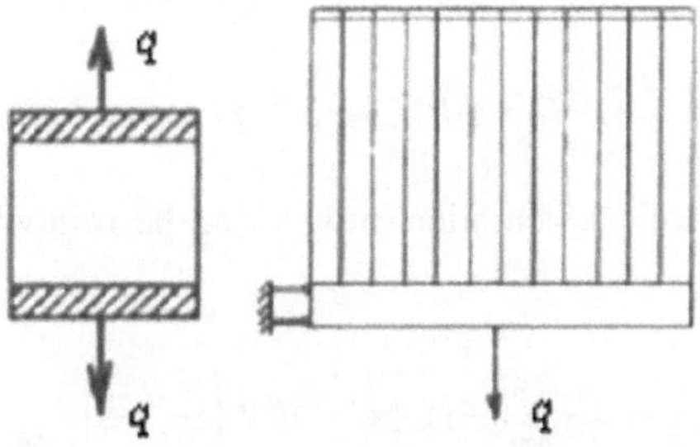

FIGURE 7.4 Parallel connection of two (left) and n (right) elements

The term $[1-P_f(0.5q)][1-P_f(0.5q)]$ on the right side of Equation (7.14) represents the probability of no failure for both bars under force of $0.5q$ on each. The term $2P_f(0.5q)[1-P_f(q)]$ is a product of the failure probability of the first bar (provided that the second bar does not fail) times the probability of no failure of the second bar (provided that the first bar fails). Coefficient 2 is introduced instead of writing down the third term which is equal to the second but can be obtained from the last with inversion of bar failure order. After manipulations the Equation (6.14) becomes:

$$P_s(q)=1-2P_f(0.5q)P_f(q)+P_f^2(0.5q) \tag{7.15}$$

This formula can be generalized (Raizer,2009) for a system of n brittle bars with parallel connection as in Figure 7.4 on the right. Based on the total probability formula, the probability of failure of i elements of n elements can be written as:

$$P_{f,i(n)}(q)=C(i,n)P_{(n-i)}(q)P_{i(n)}(q) \tag{7.16}$$

where $C(i,n)=\dfrac{n!}{i!(n-1)!}$ is the number of cases (number of combinations) favorable to the probability of failure of i elements;

$P_{(n-i)}(q)=\left[1-\dfrac{P(q)}{n-1}\right]^{n-i}$ is the probability that $(n-i)$ elements will be in a working state.

$P_{i(n)}(q)$ is a conditional probability that only i elements will fail while all other elements are remain in working state.

For a system consisting of one, two or three elements, $P_{i(n)}(q)$ is:

$$n=1,\ \ P_{0(1)}(q)=1,\ \ P_{1(1)}(q)=P(q)$$

$$n=2,\ \ P_{0(2)}(q)=1,\ \ P_{1(2)}(q)=P(q/2);P_{2(2)}(q)=-P^2(q/2)+2P(q/2)P(q)$$

$$n=3,\ \ P_{0(3)}=1,\ \ P_{1(3)}(q)=P(q/3),\ \ P_{2(3)}(q)=-P^2(q/3)P(q/2)$$

$$P_{3(3)}(q)=P^3(q/3)-3P^2(q/2)P(/3)+6P(q/3)P(q/2)P(q)$$

Applying the method of mathematical induction, the following recurrent formula is derived:

$$P_{i(n)}(q)=-\sum_{k=0}^{i-1}(-1)^{i-k}C_1^{i-k}P^{i-k}\left(\frac{q}{n-k}\right)P_{k(n)}(q) \tag{7.17}$$

It follows from Equation (7.17) that at $i = n$, $P_{f(n)} = P_{n(n)}(q)$ and the failure probability of system is:

$$P_{f,n(n)}(q) = -\sum_{k=0}^{n-1}(-1)^{n-k}C(n,n-k)P^{n-k}\left(\frac{q}{n-k}\right)P_{n(n)}(q) \tag{7.18}$$

This formula expresses the failure probability of the system with regard to the failure probability of an element. On the other side, using the same method of mathematical induction one can derive a formula for failure probability $P_{f,n(n)}(q)$, that is recurrently expressed via failure probability of partial components of the given system:

$$P_{f,n(n)}(q) = -\sum_{k=0}^{n-1}(-1)^{n-k}C(n,n-k)P^{n-k}\left(\frac{q}{n}\right)P_{k(k)}(q) \tag{7.19}$$

Systems with parallel connections are sometimes called systems with reservation. One example of parallel connection is a group of elevators in tall buildings or using a number of bolts instead of one bolt of larger diameter. Calculations in accordance with Equation (6.19) show that replacing one tensional member with three parallel ones with the same cross-section would change the probability of failure from $P_{f(1)} = 0.08076$ to $P_{f(3)} = 0.00012$.

7.4 DYNAMIC EFFECTS IN BRITTLE SYSTEMS

Dynamic effects can appear due to redistribution of stresses in the process of instantaneous rupture of brittle elements within a system (Geniev, 1993; Raizer, 2009). Let's consider a truss shown in Figure 7.5 which is statically indeterminate to the nth degree.

This truss has hinged connections in nodes; loads are applied in nodes and are proportional to parameter λ: $F_1 = \lambda f_1$, $F_2 = \lambda f_2$, ... $F_k = \lambda f_k$. If the value of λ equals to λ_m then an instantaneous rupture of the mth brittle-member makes the truss statically indeterminate to the $(n-1)^{\text{th}}$ degree (Figure 6.6). The forces in the initial system are denoted as F^s_{1n}, F^s_{2n}, ... F^s_{in}, where superscript s stands for the static character of forces. If the transition from n to $(n-1)$ in a statically indeterminate brittle system is realized in the form of very slow failure of the mth member, then forces in the $(n-1)$-member system are, respectively: $F^s_{1(n-1)}$, $F^s_{2(n-1)}$, ... $F^s_{i(n-1)}$. In case of instantaneous brittle rupture of m members of a statically indeterminate to the nth degree system, the dynamic effects are inevitable. In the new $(n-1)^{\text{th}}$ statically indeterminate system, forces are greater than $F^s_{i(n-1)}$. This is suitable with the static way of rupture. The dynamic forces can be denoted as: $F^d_{1(n-1)}$, $F^d_{2(n-1)}$, ... $F^d_{i(n-1)}$. A general analytical solution for these forces will be derived below.

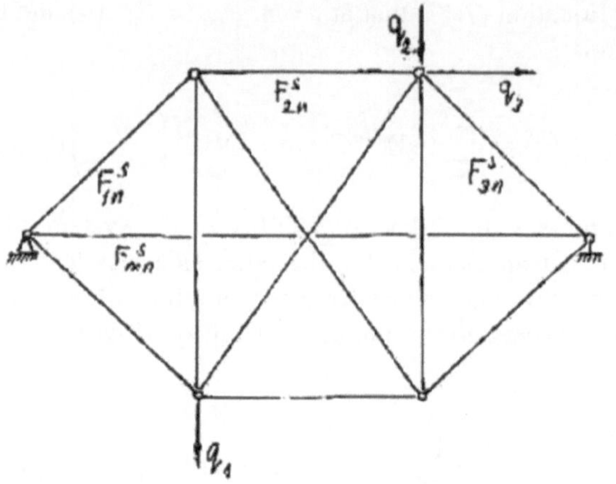

FIGURE 7.5 Statically indeterminate truss to the nth degree

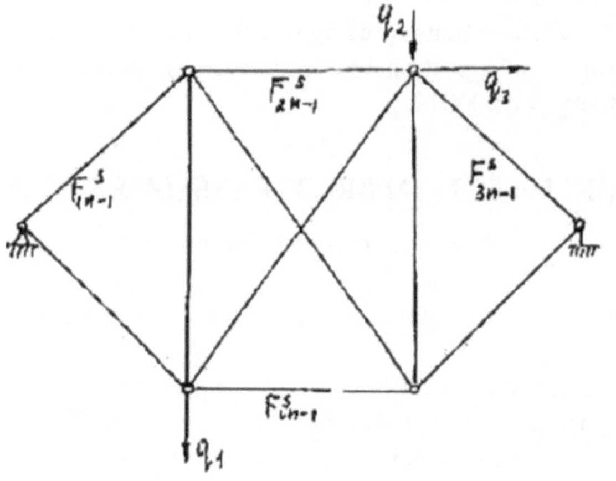

FIGURE 7.6 Truss in Figure 6.5 after the mth member failure making it statically indeterminate to the $(n–1)^{th}$ degree

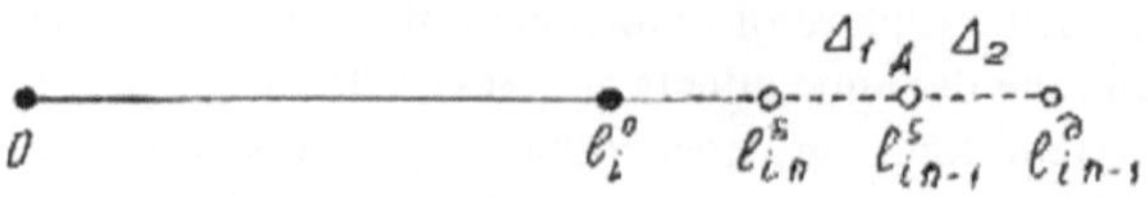

FIGURE 7.7 The ith member

Let's consider the ith member shown in Figure 6.7. The member is restrained from axial displacements at point 0 situated at one of its ends. The following designations are used:

l_i^0– length of the ith member, if force is equal to zero.
l_{in}^s, $l_{i(n-1)}^s$, $l_{i(n-1)}^d$ – length of the ith member, if force is F_{in}^s, $F_{i(n-1)}^s$, and $F_{i(n-1)}^d$, respectively, in a system ($\lambda = \lambda_m$) statically indeterminate to the nth, $(n\text{-}1)^{th}$, and $(n\text{-}1)^{th}$ degree respectively.

Corresponding variations of the ith member's length are determined by well-known relations:

$$l_{in}^s - l_i^0 = \frac{F_{in}^s}{E_i A_i} l_i^0 ; l_{i(n-1)}^s - l_i^0 = \frac{F_{i(n-1)}^s}{E_i A_i} l_i^0$$

Similarly,

$$l_{i(n-1)}^d - l_i^0 = \frac{F_{i(n-1)}^d}{E_i A_i} l_i^0$$

Hence:

$$\Delta_1 = l_{i(n-1)}^s - l_{in}^s = \frac{F_{i(n-1)}^s - F_{in}^s}{E_i A_i} l_i^0 \tag{7.20}$$

$$\Delta_2 = l_{i(n-1)}^d - l_{in}^s = \frac{F_{i(n-1)}^d - F_{in}^s}{E_i A_i} l_i^0 \tag{7.21}$$

where E is the modulus of elasticity, and A is the cross-sectional area.

Longitudinal vibrations of the ith member are induced with instantaneous rupture of the mth member. In this case the position of point A at Figure 7.7, corresponds to the length $l_{i(n-1)}^s$, and should be considered as a coordinate of the equilibrium state for the end of the ith member, while Δ_1 and Δ_2 are appropriate amplitudes.

Vibrations of longitudinal members with regard to point A of static equilibrium have a damping character since the values of amplitudes Δ_1 and Δ_2 during the first half-period can be taken, within the given margin of error, approximately equal to each other. From condition $\Delta_1 = \Delta_2$, one can get from Equations (7.20) and (7.21):

$$F_{i(n-1)}^s - F_{in}^s = F_{i(n-1)}^d - F_{i(n-1)}^s$$

hence

$$F_{i(n-1)}^d = 2F_{i(n-1)}^s - F_{in}^s \tag{7.22}$$

Equation (7.22) is valid for components both in tension and compression.

Let's consider a system organized in such a manner that the removal of the mth element from the system would cause the force in the jth element to change sign to the opposite one (under the same static load), then Equation (7.22) for dynamic force in the jth element will be as follows:

$$F^d_{i(n-1)} = 2\left|F^s_{i(n-1)}\right| - \left|F^s_{in}\right| \tag{7.23}$$

This case is rather dangerous because dynamic effects in the bar systems made of brittle materials lead to an essential increase of dynamic forces in comparison with static one. In view of equation (7.22), in the system of n members shown in Figure 7.4 in the right-hand scheme, dynamic forces in the remaining (n-1) members after instantaneous rupture of one of its bars under given load q will become:

$$F^d_{i(n-1)} = q\frac{n+1}{n(n-1)} \tag{7.24}$$

Let's introduce dynamic factor $\varphi^d_{n-1} = \dfrac{n+1}{n-1}$ as a ratio $\dfrac{F^d_{n-1}}{F^s_n}$. It becomes obvious that values of dynamic factors will increase with the number of members, n, in the initial system decreasing. Thus, when $n = 9, 7, 5, 3$ or 2, it follows that $\varphi^d_8 = 1.25$, $\varphi^d_6 = 1.33$, $\varphi^d_4 = 1.50$, $\varphi^d_2 = 2.00$, $\varphi^d_1 = 3.00$. In case of simultaneous failure of k bars out of n we have

$$F^d_{i(n-k)} = q\frac{n+k}{n(n-k)} \tag{7.25}$$

Now it is possible to generalize Equation (7.19) to take into account the effect of instantaneous brittle failure

$$P^d_{f,n(n)}(q) = -\sum_{k=0}^{n-1}(-1)^{n-k}C(n,n-k)P^{n-k}\frac{q(n+k)}{n(n-k)}P_{k,k}(q) \tag{7.26}$$

When $n = 2$, Equation (7.26) gives

$$P^d_s(q) = 1 - 2P(q/2)P(3q/2) + P^2(q/2) \tag{7.27}$$

A comparison of (7.15) and (7.27) shows that the reliability decreases when instantaneous failure occurs.

It should also be noted that systems with parallel connections could not be applied directly to reliability analysis of statically indeterminate structures. This model does not use redistribution of forces in components remaining at working state after some of them fail.

7.5 PARALLEL CONNECTION OF PLASTIC COMPONENTS

Let us consider a continuum of uniformly loaded, parallel elements such as a narrow strip of fabric fixed in a rigid frame and loaded transversally by tensile forces (Figure 6.8). The material is assumed to be plastic with the yield strength to be a stationary random function $\tilde{r}_l(x)$. The total strength of a strip of length L is a random variable

$$r = \int_0^L \tilde{r}_1(x)dx \tag{7.28}$$

The mean value of this variable is

$$\bar{r} = \int_0^L \bar{r}_1 dx = \bar{r}_1 L \tag{7.29}$$

$$\text{and variance } s_r^2 = \int_0^L\int_0^L \rho(x_1, x_2)dx_1 dx_2 \tag{7.30}$$

For the stationary function, the correlation function is:

$$\rho(x_1 - x_2) = s_{r1} \exp[-a|x_1 - x_2|] \tag{7.31}$$

where s_{r1} is the variance of the random function $\tilde{r}_1(x)$. Hence

$$s_r^2 = s_{r1}^2 \int_0^L\int_0^L \exp[-a|x_1 - x_2]|dx_1 dx_2 \tag{7.32}$$

Its integration is carried over the area of side L. Also, $\exp[-\alpha|x_1 - x_2|]$ is a function symmetric about the diagonal $x_1 = x_2$. Equation (7.32) can be transformed to:

$$s_r^2 = 2s_{r1}^2 \int_0^L e^{-\alpha t} \sqrt{2(L-t)} \frac{dt}{\sqrt{2}} \tag{7.33}$$

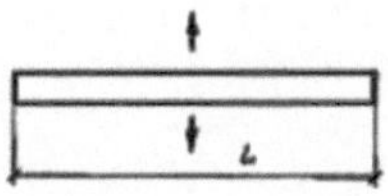

FIGURE 7.8 Strip under tension

Integrating over the area of a triangle, where $x_1 > x_2$ and $t = x_1 - x_2$ yields

$$s_r^2 = \frac{2s_{r1}^2}{\alpha^2}\left(\alpha L - 1 + e^{-\alpha L}\right) \tag{7.34}$$

For rather large L, Equation (7.34) can be simplified to

$$s_r^2 = \frac{2s_{r1}^2}{\alpha^2}(\alpha L - 1) \approx 2s_{r1}^2\frac{L}{\alpha} \tag{7.35}$$

7.6 FAILURE PROBABILITY OF MULTI-ELEMENT SYSTEMS

The failure probability of the i-th element of a system consisting from m-elements is denoted as$P_{fi}(A_i)$, and event expressing the failure of the i-th element of the system is A_i. Based on the formula for determining the probability of sum of any number of joint events, we can write it down

$$P_{fm} = \sum_{i=1}^{m} P_{fi}(A_i) - \sum_{i=1}^{m}\sum_{j=m}^{m} P_{fij}(A_i)(A_j) + \sum_{i=1}^{m}\sum_{j=1}^{m}\sum_{k=1}^{m} P_{fijk}(A_i A_j A_k) - \ldots + (-1)^m P_f(A_i, A_j, A_k, \ldots, A_l) \tag{7.36}$$

For design models of structures that are highly reliable systems and with relatively low values of the correlation coefficients of the strength of the elements ($r \le 0,5$), the common term of the alternating series (7.36) tends to be zero. In accordance with the characteristic of the Leibniz series, (7.36) is convergent. If we neglect the probability of the joint failure of more than two elements, then we can approximately assume

$$P_{fm} \approx \sum_{i=1}^{m} P_{fi}(A_i) - \sum_{i=1}^{m}\sum_{j=1}^{m} P_{fij}(A_i, A_j) \tag{7.37}$$

With one-parameter loading and the distribution of resistance according to the normal law, the probability of failure of the element is determined by the formula

$$P_{fi}(A_i) = 0,5 + \phi(\beta_i) \tag{7.38}$$

where

$$\beta_i = \frac{F_i - \bar{R}}{s_R};\ \varphi(\beta_i) = \frac{1}{\sqrt{2\pi}}\int_o^{\beta_i} e^{-\frac{t^2}{2}}dt$$

F_i – force in the element from external load;
$\bar{R}, s_R$ – mean value and standard of resistance (strength);
β_i – the reliability index of the element.

If r is the correlation coefficient between the resistances of individual elements, then, by calculating the probability of joint failure of two elements, the following form of representation is convenient (Cramér & Leadbetter, 2004):

$$P_{fij}(A_iA_j) = P_{fij}(\beta_i\beta_j r) = P_f(\beta_i)P_f(\beta_j) + \frac{1}{2\pi}\int_0^r \frac{1}{\sqrt{1-z^2}} \exp\left[\frac{\beta_i^2 - 2z\beta_i\beta_j + \beta_j^2}{2(1-z^2)}\right] dz \tag{7.39}$$

The Equation (7.39) can be reduced to the following form:

$$P_f(\beta_i\beta_j r) = P_f(\beta_i)P_f(\beta_j) + \frac{c}{\sqrt{2\pi}}\left(\arcsin r + \beta_i\beta_j\sqrt{1-r^2}\right) \tag{7.40}$$

where

$$c = \exp\left[-\left(\frac{\beta_i^2 + \beta_j^2}{2}\right)\right]$$

The Equation (7.40) has the greatest error at $r = 1$. It can be converted (Pereverzev, 1987):

$$P_f(\beta_i\beta_j r) = P_f(\beta_i)\left\{P_f(\beta_j) + \frac{\left[1 - P_f(\beta_j)\right]\left[\arcsin r + \beta_i\beta_j\left(1 - \sqrt{(1-r^2)}\right)\right]}{\frac{\pi}{2} + \beta_i\beta_j}\right\} \tag{7.41}$$

For $r = 0$ and $r = 1$, the Equation (6.41) is exact.

Table 7.1 contains the values of the probability of no failure for the system consisting of four bars, according to the model shown in Figure 7.4. For the calculation, Equations (7.37), (7.38), (7.41) were used to give equal indices of reliability and correlation coefficient.

At high reliability levels (β>3), considered correlation does not affect to the value of the probability of no failure, as it was expected.

7.7 PROBABILISTIC LIMIT EQUILIBRIUM METHOD

Elastic calculation has limited possibilities for assessing the actual load-bearing capacity, and, consequently, the reliability of structures. The assumption about

TABLE 7.1
The values of the probability of no failure for the system consisting of four bars

Reliability index	Correlation coefficient *r*					
β	0,0	0,1	0,2	0,3	0,4	0,5
1,1	0,567767	O,597975	0,631705	0,889185	0,710813	0,757278
1,3	0,669019	0,688144	0,710370	0,735900	0,765075	0,795450
1,5	0,759549	0,770773	0,784435	0,800750	0,820022	0,842709
1,7	0,833654	0,839759	0,847594	0,857375	0,869376	0,883970
1,9	0,890080	0,893160	0,897352	0,902854	0,909897	0,915783
2,1	0,930457	0,931896	0,933989	0,934893	0,940790	0,945911
2,3	0,957793	0,958417	0,959393	0,960831	0,962863	0,965656
2,5	0,975392	0,975643	0,976067	0,976736	0,977735	0,979177
2,7	0,986204	0,986297	0,986470	0,986761	0,987224	0,987929
2,9	0,992557	0,992590	0,992655	0,992774	0,992977	0,993303
3,1	0,996135	0,996145	0,996168	0,996214	0,996298	0,996440
3,3	0,998068	0,998071	0,998078	0,998079	0,999090	0,999114
3,5	0,999070	0,999071	0,999073	0,999079	0,999090	0,999114
3,7	0,999569	0,999569	0,999570	0,999571	0,999576	0,999584
3,9	0,999808	0,999808	O,999808	0,999808	0,999810	0,999813
4,1	0,999918	0,999918	0,999918	0,999918	0,999918	0,999919
4,3	0,999966	0,999966	0,999966	0,999966	0,999966	0,999966

elastic behavior of structure up to the limit state is equivalent to identifying this state with the appearance of a limit stress (yield strength, for example) at some point in the structure, that is, the assumption that the structure works on the principle of a weak link. This is true in a limited number of cases, for example for statically determinate systems. For statically indeterminate systems, the redistribution of internal forces, due to failure of sections or elements, leads in most cases to the fact that the load can increase, exceeding the level that corresponds to the appearance of yield stress, or the failure of one of the elements of a multi-element (multi-connected) system. The limit load corresponding to the transformation of the structure into a mechanism can be determined by the limit equilibrium method. The advantages of using this method in the probabilistic formulation from the standpoint of simplicity of calculation are obvious. Analysis of the reliability of ideal elastic-plastic multi-element systems can be presented as follows. It seems appropriate to consider all the mechanisms of failure that represent linear combinations of the main mechanisms.

For m possible mechanisms, the probability of failure is written as follows.

$$P_f(B) = P_f\{A_1 \cup A_2 \cup, \ldots, \cup A_m\} = 1 - P_s\left[(1-A_1)(1-A_2), \ldots, (1-A_m)\right] \quad (7.42)$$

Event B denotes the occurrence of at least one of the events A_i. The A_i event characterizes the formation of i-th mechanism. Estimates proposed for $P_f(B)$ (Cornell, 1969) are:

$$\max P_f(B) \leq P_f(B) \leq 1 - \prod_{i=1}^{m}\left[1 - P_f\left(A\ \right)\right] \tag{7.43}$$

When considering many failure mechanisms, the constraints of Equation (7.43) are quite wide, and narrower bounds are proposed in (Ditlevsen, 1979):

$$P_f(A_1) + \sum_{i=2}^{m}\left[P_f(A_i) - \sum_{j=1}^{i-1} P_f(A_i, A_j)\right] \leq P_f(B) \leq P_f(A_1) \tag{7.44}$$
$$+\sum_{i=2}^{m}\left[P_f(A_i) - \max P_f(A_i, A_j)\right]$$

Here $P_f(A_i, A_j)$ is the probability of establishing a combination mechanism of failure from the two main ones. In Equation (7.44) it is accepted that j<i.

For bars systems, the number of basic mechanisms is equal to the difference between the number of plastic hinges formed at the points of occurrence of limit moments and the degree of static indetermination. The virtual work equation can be represented as

$$W = \sum_j \theta_{ij} M_{ij} + \sum_k a_{ik} F_k \tag{7.45}$$

where F_k – denotes the load on structure; M_j and θ_{ij} – are yield moment and displacement in the j-th plastic hinge for the i-th structure failure, respectively; constitute virtual displacements.

Equation (7.45) can also be written as:

$$W = [E]\ M \tag{7.46}$$

where $[E]$ is the matrix of virtual displacements for the main mechanisms of failure; and M is the vector of limiting moments and loads.

Defining a normalized matrix $[K]$ such that its elements are

$$k_{ij} = \frac{e_{ij}}{\left(\sum e_{ij}^2\right)^{1/2}} \tag{7.47}$$

Here e_{ij} – is an element of the matrix $[E]$ (displacement of the j-th hinge with the i-th mechanism of failure). Transform the Equation (7.46):

$$W = X^T \left[K\right]^T M \tag{7.48}$$

Here X^T – is the transposable vector of weight multipliers that considers combined mechanisms (linear combinations of basic mechanisms). If we denote $D = [K]M$, then the expressions for the mean value and variance W are represented in the form

$$\bar{W} = D^T M, s^2(W) = D^T [s]^T [C] D \tag{7.49}$$

where $[s]$ is the diagonal matrix of standard deviations of loads and yield points (limiting moments); and $[C]$ is a correlation matrix whose elements represent the correlation moments between loads and between limit moments. The generalized reliability index of the system is written as

$$\beta = \bar{W} / s(W) \tag{7.50}$$

The minimum values of β correspond to the main mechanisms of failure. The problem is to determine the local minimum of β by nonlinear programming methods (Moses & Stevenson, 1970; Raiser, 1986). To calculate the probability of system failure $P_f(W)$, it is necessary to consider the statistical dependence between random forms of failure. The correlation coefficient between the two mechanisms $W_i, W_j\ (i \neq j)$ is determined by the formula

$$r_{ij} = \frac{\sum_{ij}\left[\theta_i \theta_j s^2\left(M_{ij}\right) + a_i a_j s^2\left(F_{ij}\right)\right]}{s\left(W_i\right) s\left(W_j\right)} \tag{7.51}$$

where $s^2(M_{ij})$ and $s^2(F_{ij})$ – are the variances of limiting moments and loads in the i-th plastic hinge. Consider the frame at Figure 7.9.

All loads applied to the frame F_1, F_2, F_3, F_4 (Figure 7.9) are statistically independent. It is assumed that they are distributed according to the normal law. The limiting moments M_1, M_2, M_3, M_4 depend on the yield strength and are distributed

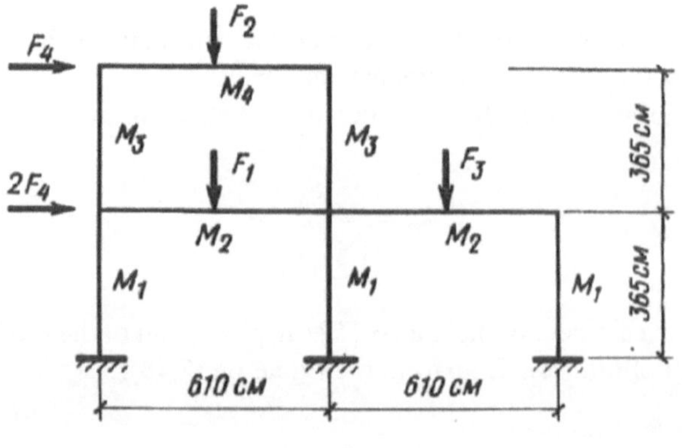

FIGURE 7.9 Model of frame

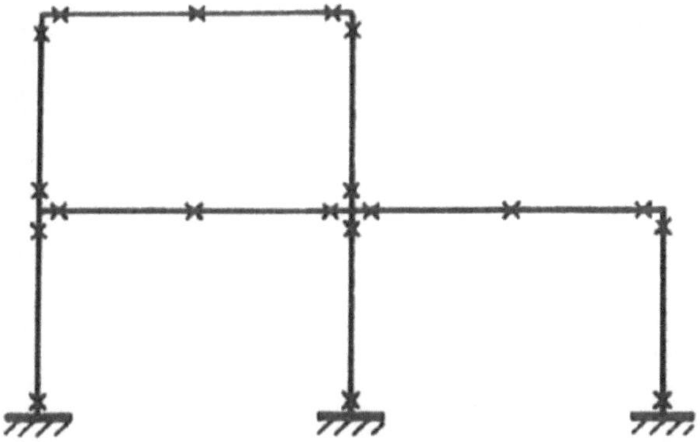

FIGURE 7.10 Potential plasticity hinges

according to the normal law. The places of their occurrence on the cross-section dimensions are also shown in Figure 6.9. It is assumed that the correlation of yield strength and cross-section dimensions for elements with the same values of limiting moments is close to functional, and therefore plasticity hinges are formed in cross-sections with maximum bending moments (Figure 7.10).

The values of F_i and M_i taken in the example are shown in Table 7.2.

The number of possible plastic hinges is 19, the degree of redundancy of the frame is 9. This implies 10 main (independent) mechanisms of failure (Figure 7.11): 3, 7, and 10 correspond to the beam failure scheme, 1 and 4 to the shear form, and the remaining mechanisms correspond to the mutual rotation of the yield joints.

In this example, 111 combinational mechanisms are possible. Of these, 12 identified main mechanisms (Figure 6.12) will correspond to the local minimum of the reliability index and the highest values of the probability for formation of the i-th mechanism (Table 7.3).

It can be noted that the value of the reliability index for all main mechanisms varies in a narrow range between 1,854 and 2,174. The upper and lower bound for the probability of system failure, calculated by the Equation (7.44), is

$$0{,}032 \le P_f(W) \le 0{,}265 \tag{7.52}$$

The lower estimate coincides with the probability of failure with the formation of the first mechanism according to Table 7.3. Assuming that the failure forms are correlated if $r_{ij} \ge r_0$, where r_{ij} is defined by the Equation (7.51), and r_o is set to 0.7 or 0.8, it is shown that mechanisms 3, 4, 7, 9, 10, and 11 are represented by the 1st mechanism, the 12th mechanism by the 5th mechanism. Then the probability of failure of the system will be

$$\begin{aligned} P_f(W) = {} & P_f(W_1 < 0) + P_f(W_2 < 0) + P_f(W_5 < 0) + P_f(W_6 < 0) \\ & + P_f(W_8 < 0) = 0{,}128 \end{aligned} \tag{7.53}$$

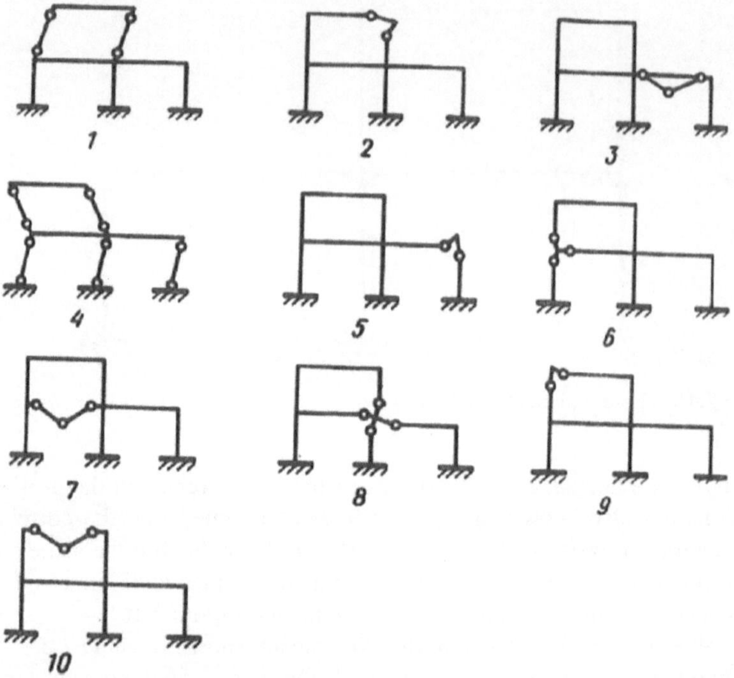

FIGURE 7.11 Main (independent) mechanisms of failure

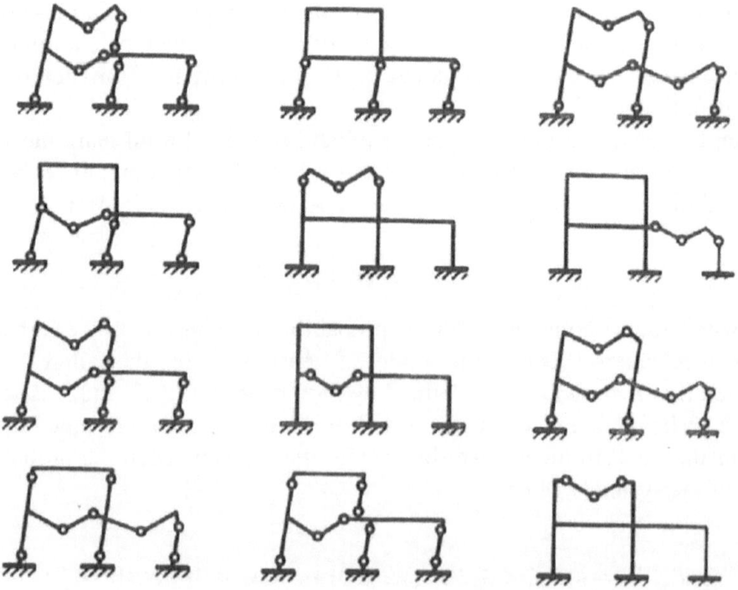

FIGURE 7.12 The most likely mechanisms of failure

TABLE 7.2
The values of F and M

Values of limit moments and loads	Mean values кНм	Variation coefficient
M_1	95	0,15
M_2	204	0,15
M_3	122	0,15
M_4	163	0,15
F_1	169	0,15
F_2	189	0,25
F_3	116	0,25
F_4	31	0,25

TABLE 7.3
The highest values of probability for i-th mechanism

Mechanism	β_i	$P_f(W_i < 0)$
1	1,854	$31,9 \cdot 10^{-3}$
2	1,886	$29,7 \cdot 10^{-3}$
3	1,917	$27,8 \cdot 10^{-3}$
4	1,967	$24,6 \cdot 10^{-3}$
5	1,981	$23,8 \cdot 10^{-3}$
6	1,996	$23,0 \cdot 10^{-3}$
7	2,064	$19,6 \cdot 10^{-3}$
8	2,065	$19,5 \cdot 10^{-3}$
9	2,088	$18,4 \cdot 10^{-3}$
10	2,149	$15,9 \cdot 10^{-3}$
11	2,155	$15,6 \cdot 10^{-3}$
12	2,174	$14,9 \cdot 10^{-3}$

The values of the lower and upper bounds according to Equation (7.44) were

$$0,071 \le P_f(W) \le 0,187 \tag{7.54}$$

The results presented here demonstrate possible approaches for analysis of the reliability of complex multi-element systems.

8 Loads and Their Combinations

Design of buildings and structures against multiple time-varying loads has always been a challenge to engineers. At most loadings on structures are stochastic in nature, an overriding concern has been proper treatment of the uncertainty such that the structural performance will be satisfactory under the loadings during its lifetime. It requires a careful consideration of the difficult load combination problem and yet the design criteria have to be simple for implementation in codes and standards.

Yi Kwei Wen, 1993

There are two kinds of statistics, the kind you look up, and the kind you make up.

RexStout

8.1 CLASSIFICATION

The action of loads on a structure is determined by the environment, more specifically, by natural factors such as snow, wind, air temperature, solar radiation, earthquake, and others, as well as factors such as human activities including live loads, gravity, movements of transport, and so on. It should be noted that the values and other load parameters considered in design become associated with the loading model but not with the real process of loading. Meaningful physical analysis should precede stochastic modeling of loads. It should include:

- load variability associated with the climatic and technology features,
- spatial and temporal variability of loads,
- depending on conditions of their occurrence and grouping these conditions along with the regions, type, and form of structures and procedures of their operations,
- observation and measurement methods with estimates of homogeneity and stability of the observation conditions and presentation of the results,
- methods to process and arrange the results of measurements.

A probabilistic model of the load is formulated on the basis of that analysis. Loads vary greatly. Mathematically, loads can be presented as random values, random

DOI: 10.1201/9781003265993-8

functions of time, as well as variable in time and space according to stochastic and deterministic laws as evidenced in works by Augusti et al. (1984), Bendat and Piersol (1971), Gnedenko et al. (1969), Gumbel (1967), Pereverzev (1987), Raizer (1995, 2009), Rzhanitsyn (1978), Smith (1986), Spaethe (1987), Sveshnikov (1968), and Tikhonov (1970). According to the ultimate limit state for estimating structural reliability, only the maximum values of loads for the considered period of time are important. The stationary process is quite satisfactory for describing the loading on structure while the maximum can be taken as statistically independent. Then it is possible to divide the service life of the structure, T, into large intervals τ. The maximums in the neighboring intervals can be considered statistically independent. The loading process can also be considered as a testing process with the number of tests equal $n = T / \tau$. In each interval τ, the loads are given as distributions of their random maximums. (It should be noted that for some actions their maximum value can have upper limits. For example, pressure of liquids in open reservoirs cannot exceed its full crowding, and the lifting capacity of a crane is also limited.)

For atmospheric actions this interval can be equal to one year. The distribution of maximums is understood as a distribution of annual maximums while the number of tests n is understood as the number of years of the structure's service life. The loads have often been presented in the form of a stationary sequence. Characteristics of natural effects are determined by environmental conditions. In reality, actions that occur simultaneously are often statistically dependent to a certain extent; typical cases are wind, snow, temperature, and so on. For simplification, each single load can be considered statistically independent in time and space. For atmospheric actions the initial statistical information has been presented in the meteorological reference sources. These considerations make the basis for formulating the above-mentioned probabilistic approach. For example, snow load is presented as a random process in the design of structures with creep deformations. The Gumbel distribution is often considered for the ultimate limit state.

Based on observations, the normal Poisson distribution has been used for the live loads. The Gauss distribution function was recommended for gravity of structures and stored materials. These and other probabilistic models for different kinds of loads are necessary for the probability-based design (Augusti et al., 1984; Raizer, 2009; Spaethe, 1987).

There are some types of classifications of loads and actions. The actions are classed according to their variations in time as follows:

- Permanent actions, whose variations in time are mostly rare, small, and slow, are assumed to remain constant in time, and the statistical distribution function varies among similar structures.
- Variable actions, whose variations in time are frequent, have their statistical distribution function varying among similar structures and in time.
- Accidental actions are actions, which occur in accidental situations or similar unpredictable cases.

According to action variation in space they are divided into two groups:

- Fixed actions, whose distribution over the structure is unambiguously defined by one parameter, while the magnitude of a particular action may change.
- Free actions, which may have any arbitrary distribution over the structure within given limits.

Actions, which cannot be assigned to either of these two groups, are considered to consist of a fixed part and a free part.

According to the way in which the action is acting on the structure, a distinction is made between:

- static actions, which do not cause significant accelerations of the structure or a structural member,
- dynamic actions, which cause significant accelerations of the structure.

The need to account for the dynamic character of load is governed mainly by the structure's properties. In design practice the dynamic effect is usually represented by a dynamic factor applied to the static load.

Actions, which are creating forces, stresses, deformations, and displacements in structure, could be called direct actions. Indirect actions are, for example, biological (e.g., decay of wooden structures), chemical (e.g. corrosion effect), and thermal (including fire). These actions influence the load-bearing capacity and structural durability. The type of structural materials used also determines the necessity to analyze the structure for its resistance to deep freeze, humidity, and other indirect actions.

Structures are usually subjected to several loads. The probability of several loads acting simultaneously is usually considerably less than that for either of them provided that the loads are of comparable value. Therefore, special combination factors for combinations of loads and other approaches for determination of combination factor are used in many national and international regulations, such as Ferry-Castanheta (1971) model, Turkstra rule (1970), and design value method (ST 2394, 2015]).

8.2 STATISTICAL MODELS OF VARIABLE ACTIONS

The main qualitative characteristic of actions is their intensiveness as a function of space coordinates and time. Probabilistic qualities of the intensiveness as a random variable are evaluated by statistical sampling for random values and by sampling functions for random functions (realizations). Main types of initial realizations of measured loads are shown in Figure 8.1:

- Differentiable – Figure 8.1a, b.
- Non-differentiable (multi graded type):
  - o with connected pulses – Figure 8.1c.
  - o with unconnected pulses – Figure 8.1d–h, including instant pulses,
  - o Figure 8.1f, h.

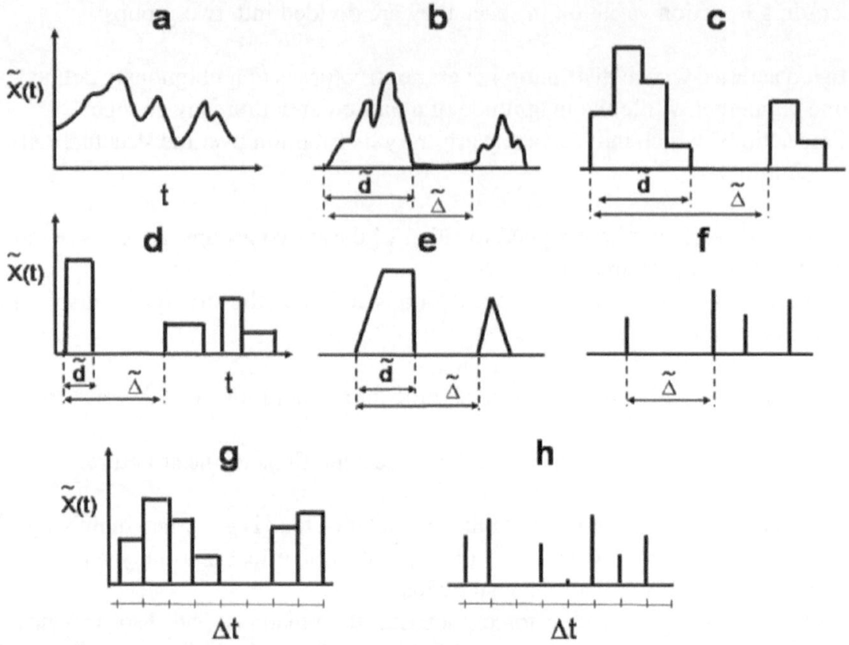

FIGURE 8.1 Types of measured loads

Each pulse starts with zero intensiveness. Sequences of the loading in the "b-f" realizations, caused by a random flow of events, consist of appearance of non-zero pulse lasting for a period of $\tilde{d} \le \tilde{\Delta}$ where $\tilde{\Delta}$ is the random interval of return period. Stationary flows of many load occurrences, such as crane and transport loads or loads due to machinery operation in buildings, can be simulated by the Poisson flow model with corresponding exponential distribution:

$$F(t) = 1 - \exp(-\overline{\nu} t), \quad t \ge 0, \quad \overline{\nu} > 0 \tag{8.1}$$

Here $\overline{\Delta} = \frac{1}{\overline{\nu}}$, $s_{\Delta} = \frac{1}{\overline{\nu}}$, and $\overline{\nu}$ is the intensiveness (density) of flow or the mean frequency of event occurrence. For space realizations the interval $\overline{\Delta}$ is considered as an element of length, area, or volume. The random number of flow events N_{τ} in τ interval is described by Poisson distribution:

$$P_N(t) = P\left[\tilde{N}_{\tau} = N\right] = \frac{(\overline{\nu}\tau)^N}{N!} e^{-\overline{\nu\tau}} \tag{8.2}$$

Parameters are $\overline{N}_t = \overline{\nu}\tau = \frac{\tau}{\overline{\Delta}}$; $s^2_{N_\tau} = \overline{\nu\tau}$. The probability that events will not appear in the interval τ is:

$$P_0(t) = P[\tilde{N}_\tau = 0] = e^{-\nu t} \tag{8.3}$$

The time-depending Poissson flow with variable instant density $\nu(t)$ can be replaced at a chosen interval τ beginning from t_j with a stationary flow of an equivalent mean density:

$$\bar{\nu}(t_j; t_j + \tau) = \frac{1}{\tau} \int_{t_j}^{t_j+\tau} \nu(t) dt \tag{8.4}$$

The temporal pattern of realization shown at Figures 8.1g and h is determined not only by characteristics of real processes but by the regular flow of measurements performed at fixed frequency λ (e.g. for meteorological variables $\lambda = 4, 8, 24$ times per day) and with deterministic interval $\Delta_t = \frac{1}{\lambda}$.

If the load duration process is ignored (no serviceability limit state considered), the process of loading can be presented as a realization of instant pulses of discreet time (random sequence)—Figure 8.1h. Intensiveness of the non-rectangular pulse (Figure 8.1b, c, e) is characterized by instant values of $\tilde{X}$(t) and the maximum $\tilde{X}_\Delta$. In the rectangular pulse (Figure 8.1d, g) all the instant values are equal to the maximum one. In loading "b-e" (Figure 8.1), modeled by a Poisson flow with the frequency of impulses equal to $\bar{\nu} = \frac{1}{\Delta}$, if only the maximum values of intensiveness of each impulse $\tilde{X}_\Delta$ are chosen, then these values became the realization in Figure 8.1f. Basic probabilistic features of the load intensiveness $\tilde{X}$(t) are characterized by its one-dimensional distribution function $F_x(x)$. This function is determined as the probability of the event $\tilde{X} < x$ at a time instant, that is, $F_x(x) = P(\tilde{X} < x)$. The supplement of the distribution function to 1 (distribution at a tail) determines the probability of the opposite event $\tilde{X} \geq x$:

$$G_x(x) = 1 - F_x(x) = P(\tilde{X} \geq x) \tag{8.5}$$

For the stationary processes, such as "a-e" in Figure 7.1, the one-dimensional statistical distribution function (statistical estimate) $F_x^0(x)$ should be defined as the relative duration for the considered process to remain below the level of "x"—see Figure 8.2.

$$F_x^0(x) = 1 - G_x^0(x) = 1 - \frac{\sum^{n_x} \alpha_i \left[\tilde{X}(t) \geq x\right]}{T} \tag{8.6}$$

where T is the total duration of realization, n_x is the number of $\tilde{X}(x) \geq x$ defined at interval T (to be calculated), and $\alpha_i \left[\tilde{X}(t) \geq 0\right]$ is the persistent duration.

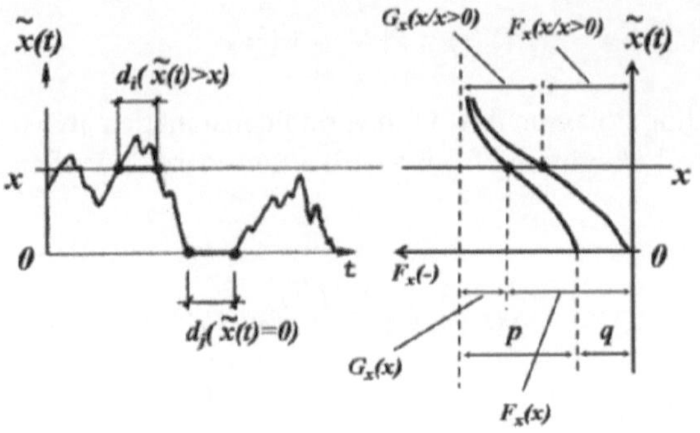

FIGURE 8.2 Process of loading depending on time

It should be noted that the stationary random processes being not decomposable to incompatible processes of different types and being also observed in homogeneous and time-stable conditions (in real place and environment), or under stable technological conditions, have, as a rule, ergodic features with regard to mean value and correlation function.

For the time-dependent processes (Figure 8.2) the statistical distribution function is defined by processing the ensemble of m realizations $x_i(t)$ of duration T_k. This procedure is taken by equidistant sections of the process, which corresponds to fixed time instants t_r $(r = 1,2,...,k)$. At that each instant t_r defines the middle of the interval:

$$\Delta_k = \frac{T_k}{k}$$

The typical interval of transient effect T_k can be determined from some preliminary analysis of the considered process. For climatic variables, for example, the periodical transient effect is associated with a period of one year, and sometimes a period of 24 hours. When $t = t_r$, from the ensemble of realizations the statistical sampling performs with m volume as follows:

$$F_x^0(x/t_r) = 1 - G_x^0(x/t_r) = 1 - \frac{n_{x,m}(t_r)}{m} \tag{8.7}$$

where $n_{x,m}(t_r)$ is a number of outliers $\tilde{X}(t_r) \geq x$ in the sampling:

$$\overline{X^0}(t_r) = \frac{1}{m}\sum_{i=1}^{m} x_i(t_r);\ s_x^{0^2}(t_r) = \frac{1}{m-1}\sum_{i=1}^{m}\left[x_i(t_r) - \overline{x^0}(t_r)\right]^2$$

Realizations of climatic variables with transient effect, like years, permit to consider and to treat them as conditionally stationary. In this case X^0 is an estimate of the mean value of the process during many years and $s_x^{0^2}$ includes an essential part of deviation of a year harmonic $\overline{X^0}(t)$ relative to $\overline{X^0}$.

8.3 CLIMATIC ACTIONS ON STRUCTURES

Turning directly to the consideration of climatic loads (snow, wind, temperature) on structures, it can be noted the following.

The intensity of snow load on the building roof can be represented in general terms as a random function of coordinates (*x, y*), time *t*, terrain factors *g*, and structure factors *C*:

$$\tilde{q}(x,y,t/g,C)=\tilde{q}_0(t/g)\tilde{k}(g,C/q_0)\tilde{\mu}(x,y/g,C,q_0) \tag{8.8}$$

where

$$\tilde{q}_0(t/g)-$$

is the main climatic parameter of snow load, which characterizes the real sum of solid and liquid precipitation accumulated by time *t*, falling from the atmosphere in a given area—water supply in the snow cover (in mm), weight of the snow cover (kPa) on the horizontal surface of the earth.

$\tilde{k}(g,C/q_0)-$ is the function of the transition from the weight of the snow cover of the earth $\tilde{q}_0$ to the average intensity of the snow load on the roof surface;

$\tilde{\mu}(x,y/g,C,q_0)-$ is the function of redistribution of snow load on the roof (function of the load scheme).

The factors of terrain *g* include its geographical location, topography, and roughness, which determine the nature and intensity of snow accumulation on the ground, wind, snow-wind, and temperature-humidity conditions of the terrain, the protection of the weather station and structures. Factors of structure *C* include its shape, size, orientation, surface roughness, internal temperature regime, and thermal properties of the structural coverage, which determine the snow load on the coverage in interaction with terrain factors.

The functions $\tilde{k}(g,C/q_0),\tilde{\mu}(x,y/g,C,q_0)$ are extremely complex for statistical study and probabilistic modeling and are reduced mainly to individual special cases, certain forms of building roof, and specific local conditions, and end with the development of specific recommendations for snow load schemes and transition coefficients.

When analyzing structures for the first group of limit states, the load-bearing capacity requires knowledge of the maximum snow load for many years. Since there is practically no correlation between annual maxima, the long-term snow load can be obtained theoretically by knowing the distribution function of the maximum annual load. The probability density distribution curve for the maximum snow load in one year $f(q_{01})$ is not symmetric, and $f(q_{0n})$ cannot be less than zero. The variability is quite high and the double exponential law of the integral distribution, the Gumbel distribution, can be used to approximate it (Gumbel, 1967):

$$F(q_{01}) = \exp\left(-\exp\frac{a - q_{01}}{b}\right) \tag{8.9}$$

The probability density function will be

$$f(q_{01}) = \frac{1}{b}\exp\left(-\exp\frac{a - q_{01}}{b} + \frac{a - q_{01}}{b}\right) \tag{8.10}$$
$$-\infty < a < \infty, b > 0$$

Parameters a and b are related to expectation $\overline{q}_{01}$ and variance $s^2(q_{01})$ as follows

$$\overline{q}_{01} = a + 0,5776b, \tag{8.11}$$
$$s^2(q_{01}) = 1,645b^2$$

The coefficients a and b have different values for different areas. To determine the maximum permissible snow load on a structure designed for n years, one can write down the probability of not exceeding its value as

$$F(q_{01}) = F^n(q_{01}) \tag{8.12}$$

Denoting by V the given probability of exceeding the value q_{0n} for n years, it follows

$$V = 1 - F^n(q_{01}); n = \frac{\log(1 - V)}{\log F(q_{01})}; \tag{8.13}$$
$$F(q_{01}) = \sqrt[n]{1 - V}$$

Applying Equation (8.12) to the distribution (7.9), it can be re-written as follows:

$$F(q_{0n}) = \exp\left(-\exp\frac{a_n - q_{01}}{b}\right) \tag{8.14}$$

where $a_n = a + b\ln n$.

For the mathematical expectation, one has

$$\overline{q}_{0n} = \overline{q}_{01} + \ln n \tag{8.15}$$

The variances do not change $s^2(q_{0n}) = s^2(q_{01})$

When considering structures for the second group of limit states, and analyzing deformations, it is necessary to consider the random process of snow accumulation.

The wind load on the structure is defined as the sum of the average and pulsation components and is calculated through the intensity of wind pressure:

$$W = 0.6V(t) \tag{8.16}$$

where *V(t)* is wind speed (м/sec.) is the main factor of wind load, which depends on the geographical location of the area and on associated atmospheric disturbances (cyclones, storms, thunderstorms), and also on the topographic features of the distribution of wind flows and the wind regime, as well as the roughness of the area (the nature of the surface, vegetation, obstacles, buildings and structures). The processing and smoothing of the urgent values of the wind velocity modulus for many weather stations shows that the best agreement with the statistical data has an unbiased Weibull distribution (Raizer, 2010):

$$F(V) = 1 = \exp\left[-\left(\frac{V}{\eta}\right)^{\alpha}\right] \tag{8.17}$$

where *F(V)* is the probability that the wind speed will not exceed the value of *V* at a predetermined time; α, η are coefficients, which will be determined for each weather station and depending on the wind regime for the given area.

The estimation of the parameters can be obtained by the maximum likelihood method using a sample of daily maxima of wind speeds.

Structures of buildings and open structures are also exposed to air temperature and solar radiation during operation, which cause changes in the temperature of structural elements, deformations, and displacements. Temperature forces may occur in statically indeterminate systems. Changes in air temperature over time are a random process in which two periodic fluctuations are clearly distinguished: with annual and daily periods.

In application to building structures analysis, temperature changes over time can be represented as the sum of periodic temperature fluctuations with a period equal to one year (seasonal fluctuations: summer-winter) with a random amplitude; periodic temperature fluctuations with a period equal to one day (daily fluctuations: day-night) and also with a random amplitude; and non-periodic temperature fluctuations at time intervals of several days.

$$\tilde{T}_{+}(t) = \tilde{\tau}_{k+}(t) - \tilde{\tau}^{0}_{k-}, \tilde{T} = \tilde{\tau}_{k-}(t) - \tilde{\tau}^{0}_{k+} \tag{8.18}$$

Only those increments that are caused by changes in the average daily outdoor air temperature over time are considered $\tilde{\tau}_{1+}(t), \tilde{\tau}_{1-}(t)$ that is,

$$\tilde{T}_{+}(t) \cong \tilde{\tau}_{1+}(t) - \tilde{\tau}^{0}_{-}; \tilde{T}_{-}(t) \cong \tilde{\tau}_{1-}(t) - \tilde{\tau}^{0}_{+} \tag{8.19}$$

The results of successive measurements of the average daily air temperature over time represent an implementation determined on a regular stream of observations averaged over each calendar day. The most significant feature of the random process of average daily temperature is periodic transient, which manifests itself in the annual course of temperature, its distribution function, density, and parameters. At the same time, the variability of the average daily air temperature in the winter months is much greater than in the summer months, and the ratio of standards $s(\tau_{1-})/s(\tau_{1+})$ can reach 3 … 4. Data on the repeatability of the average daily air temperature in certain months of the year makes it possible to approximate the distribution function $F_{\tau_1}(\tau/t)$, density $f_{\tau_1}(\tau/t)$, and $\overline{\tau}_1(t), s_{\tau_1}(t)$. Analysis of this data for many weather stations shows that the density function is close to a normal one,

$$f_{\tau_1}(t) = \frac{1}{s_{\tau_1}\sqrt{2\pi}} \exp\left\{ -\frac{\left[\tau - \overline{\tau}_1(t)\right]}{2s_{\tau_1}^2(t)} \right\} \tag{8.20}$$

Data on day-to-day variability allow us to estimate the derivative $\tilde{V}_{\tau_1}(t)$ of the average daily temperature process and bring the initial implementations to conditionally differentiable ones. The distribution function $\tilde{V}_{\tau_1}(t)$ in individual months of the year is close to normal with zero mathematical expectation, and the values of the average day-to-day variability are an estimate of the conditional mathematical expectation $\overline{V}_{\tau_1}^{+}(t)$positive or $\overline{V}_{\tau_1}^{-}(t)$ negative values of the derivative in certain months of the year.

8.4 CONSIDERATION OF RANDOM WIND SPEED DIRECTIONS

Wind flow is characterized not only by a random change in the speed value, but also by a random change in direction. In meteorology, wind direction is fixed by eight bearings. Primary meteorological sources contain data on the frequency of winds of different bearings, which can be used to build a wind rose for different areas. For example, the average annual long-term wind rose for Moscow, Russia is shown in Figure 8.3. From this figure it follows that the prevailing wind directions are west and south-west.

Two methods can be used taking into account random changes of the wind direction. The first method is based on the assumption that all horizontal wind

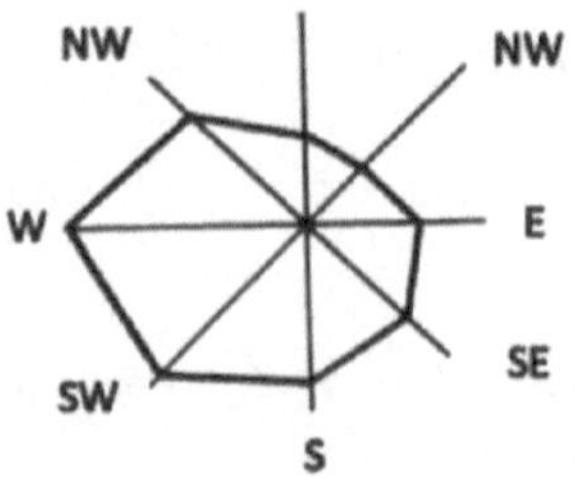

FIGURE 8.3 Wind rose

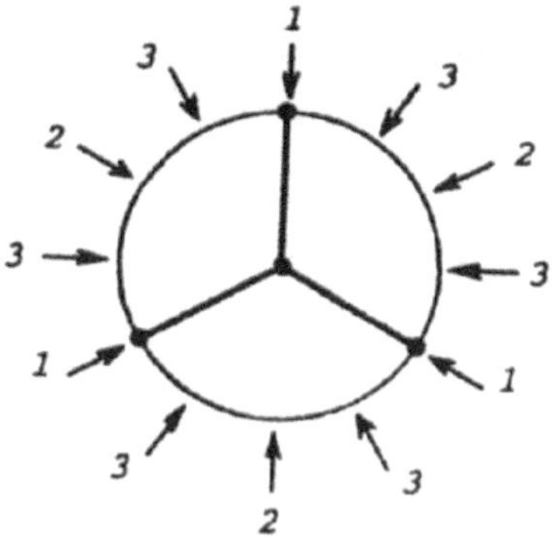

FIGURE 8.4 Diagram of the mast tie-off unit

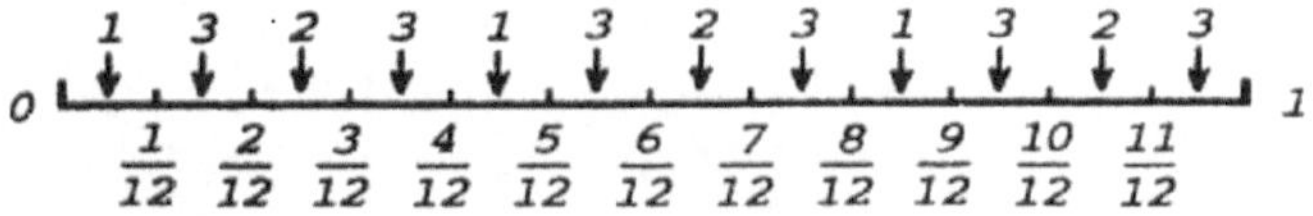

FIGURE 8.5 Scan the circle

directions are equally probable. The assumption that the maximum wind directions are equally probable is acceptable because the structure is usually oriented in space not according to the wind rose, but in accordance with the accepted architectural and planning decision, taking into account the terrain, and so on. The application of this method can be considered in the example of a mast that has three guy wires in each guy node (Figure 8.4). The nodal load is moved around the circle from the drawbar with the number 1 clockwise and the numbers of calculated cases that will occur are recorded.

As a result, a sequence of 12 terms is obtained, the meaning of which is the number of the calculated case:

$$N_d = 1,3,2;3,1,3;2,3,1;3,2,3 \tag{8.21}$$

Thus, 12 calculated cases can be specified on the circle. The number of cases of type 1 is three, type 2 is also three, and type 3 is six. The number of the calculated case can be represented as a discrete random variable taking the values 1, 2, 3 with the probability $F_{N,d=1} = 0,25; F_{N,d=2} = 0,25; F_{N,d=3} = 0,50$.

The discrete random variable is modeled by the Monte Carlo method (Shinozuka & Deodatis, 1991, 1996; Spanos & Zeldin, 1998; Elishakoff, 2003).

For this the circle Figure 8.4 unfolds into a segment that is divided into 12 equal parts (Figure 8.5) and having boundaries $|0, 1|$.

Then a sequence of random numbers ξ_i evenly distributed over the interval $|0,1|$ is generated according to which of the 12 parts of this segment the next

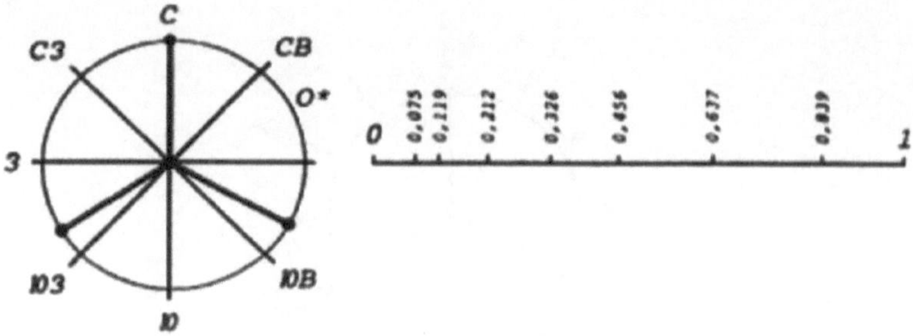

FIGURE 8.6 Random character of the wind direction

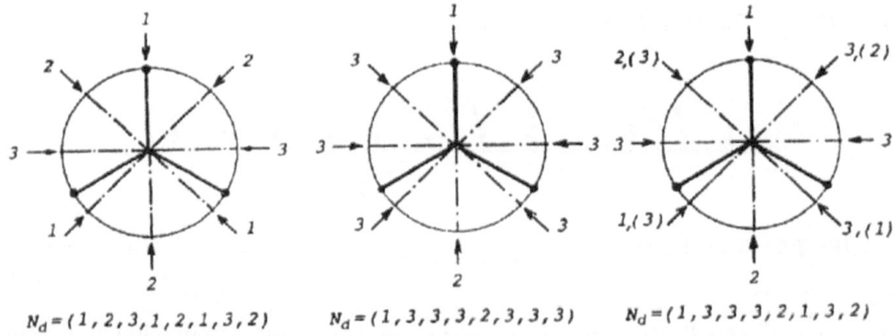

FIGURE 8.7 Accounting of random wind direction

developed number will fall, the number of the calculation case is selected, for which it is necessary to make a deterministic calculation. For example, a random number $\xi_i = 0{,}513$ will fall in the range from 6/12 to 7/12, which corresponds to the number of the calculated case $N_d = 2$. The number of intervals corresponding to the calculated case of type 1 is three, type 2 is three, and type 3 is six. Hence, as noted earlier, $F_{N,d=1} = 0{,}25; F_{N,d=2} = 0{,}25; F_{N,d=3} = 0{,}50$.

The difference between the second method and the first is that the wind rose is used instead of hypothesis of equal probability of wind directions. The circle 0^* unfolds into a segment with boundaries $|0, 1|$, which is divided into eight sections (Figure 8.6) by the number of bearings.

If one moves along the wind rose (Figure 8.3) from the north direction clockwise, then the length of the first section corresponding to the north wind direction will be numerically equal to the repeatability of this direction, expressed in relative units (Figure 8. 6). The same is done with the other sections of the segment, which has a length equal to the repeatability of the corresponding wind direction.

By virtue of the fact that there is a discrepancy between the direction of the braces in the plan and the wind directions (Figure 8.7) (with four delays in terms of such discrepancies), four variants of the sequence of eight terms are obtained, the meaning of which is the number of the designed case.

TABLE 8.1
Values of the probability for design cases

Cases	F_{N_d}-a	F_{N_d}-b	F_{N_d}-c	F_{N_d}-d
1	0,370	0,075	0,256	0,189
2	0.336	0,130	0,474	0,194
3	0,294	0,795	0,270	0,617

The values of the probabilities of occurrence for designed cases 1, 2, 3 for each of the four variants of the sequences, considering the real wind rose, are presented in Table 8.1.

- *a.* $N_d = 1,2,3,1,2,1,3,2$
- *b.* $N_d = 1,3,3,3,2,3,3,3$ *(7.22)*
- *c.* $N_d = 1,3,3,3,2,1.3,2$
- *d.* $N_d = 1,2,3,1,2,3,3,3$

8.5 LOADS ON BUILDING FLOORS

Probabilistic modeling of technological loads has features in comparison with climatic actions. Due to the exceptional diversity of functional and technological conditions implemented in buildings with different space-planning decisions and structural schemes, the possibilities of the statistical approach for direct standardizing of loads are sharply limited, and the change in technological conditions over time, which is expressed in the redevelopment and replacement of equipment, the prospective growth of loads, prevents the direct transfer of statistical data to future conditions.

The presence of internal technologically determined relationships in the implementations and samples of technological loads makes it difficult to interpret the probabilistic results of observations and selection of appropriate distribution functions of the intensity of loads and it also imposes significant restrictions on the formation of homogeneous statistical populations based on technological characteristics. Statistical estimates refer to homogeneous spatial assembly, as a rule (for example, formed from zones of influence of the same type of structural elements of one or more production sites), and they are related to the arbitrariness of the selection of an element from such an assembly. (The zone of influence of a structural element is considered to be a part of the floor area, the loading of which by an arbitrarily located concentrated force causes displacement and internal forces in the element under consideration.) The outline of the zone of influence is usually determined in the assumption of a beam structural diagram with a hinged support of the elements. The zone of influence is the scope of surface influence for the effect of one kind or another—the pressure on column, bending moment at mid-span gird, and so on. On this basis, the characteristics of the spatial variability of the intensity of technological loads as random variables are introduced, although this variability is due to a deterministic technological decision. Probabilistic modeling of technological loads can be partly justified by the fact that the construction design of buildings is often carried out in conditions of technological uncertainty, in the absence of a specific technological zoning and any data on its possible changes during the normal operation of the building.

The probabilistic model of static loads from technological equipment can be based on the following initial data:

- a standard set of elements of the main technological equipment for a production site of a certain purpose, indicating the weight (mass) equipment type «*j*», Q_j, the area of its size Δ_j, the relative frequency of its use v_j in this area;
- relative density of equipment placement on the site α_Q, estimated share of the area occupied by the equipment (possible on the rest of the area $\alpha_0 = 1 - \alpha_Q$ loads from people, auxiliary materials, individual vehicles are not considered here and should be taken into account additionally).

Loading the area of the zone of influence of a structural element is considered as the implementation of a Poisson event flow, where an event is understood as the appearance of an element of equipment in a certain unit spot with a random area (because in a typical set of areas, the size of the equipment differs). Mathematical expectation of the cell area is:

$$\bar{\Delta} = \sum_j v_j \bar{\Delta}_j \qquad (8.23)$$

where v_j can be used as an estimate of the conditional probability of occurrence of an item of equipment of type *j* in an arbitrarily selected cell (provided that the cell is loaded and is not free and assuming that the value of v_j in terms of area A_φ will not change, although the density of its distribution may, in principle, be uneven). The average number of cells' area A_φ, in which the equipment can potentially appear, will be

$$\bar{n}_A = \frac{A_\varphi}{\bar{\Delta}} \qquad (8.24)$$

Relative density of equipment location $\bar{\alpha}_Q$ is entered as an average estimation of the probability of occurrence of an item of equipment in an arbitrary unit cell for a given production site.

Then the loading of the zone of influence of the structural element is considered as the result $n_A = \bar{n}_A$ of independent tests according to a simple Bernoulli scheme with probability of occurrence:

α_Q

and probability of non-occurrence:

$$(1-\alpha_Q)$$

an item of equipment in one test (single cell).

The number of loaded cells of the zone of influence A_φ is a discrete random variable with a binomial distribution $\tilde{n}_{QA}$ and parameters:

$$\bar{n}_{QA} = \bar{\alpha}_Q n_A, S^2_{nQA} = \bar{\alpha}_Q(1-\bar{\alpha}_Q)n_A \tag{8.25}$$

This means that the relative density of equipment location in an arbitrary zone of influence is also a random variable $\tilde{\alpha}_{QA} = \frac{\tilde{n}_{QA}}{n_A}$ with parameters:

$$\bar{\alpha}_{QA} = \frac{\bar{n}_{QA}}{n_A} = \bar{\alpha}_Q, S^2_{\alpha_{QA}} = \frac{S^2_{n_{QA}}}{n_A^2} = \frac{\bar{\alpha}_Q(1-\bar{\alpha}_Q)}{n_A} \tag{8.26}$$

Applicability of the simple Bernoulli scheme as a model for estimating the spatial variability $\tilde{\alpha}_{QA}$ was confirmed by the analysis of planning schemes of suspended beltline systems of automobile plants and of the results of the survey of loads from furniture in the basic premises of office blocks (Raizer, 2010).

Value n^*_{QA} is defined as the root of the equation:

$$1-\omega = \sum_{m=n^*_{QA}}^{n_A} \binom{n_A}{m} \bar{\alpha}_Q^m (1-\bar{\alpha}_Q)^{n_A-m} \tag{8.27}$$

obtained by the selection of a suitable whole number n^*_{QA}, starting by brute force method from n_A with a sequential decrease by one. Appropriate ω -quintile is:

$$\alpha^*_{QA} = \frac{n^*_{QA}}{n_A}$$

The intensity of the loads of equipment Q in an arbitrary unit cell is defined as the result of one of the incompatible events of load occurrence Q_j with probability v_j, forming a complete group $\left(\sum_j v_j = 1\right)$:

$$\tilde{Q} = \sum_j v_j \tilde{Q}_j; \bar{Q} = \sum_j v_j \bar{Q}_j; s^2_Q = \sum_j v_j \left[s^2_{Q_j} + \left(\bar{Q} - \bar{Q}_j\right)\right]^2 \tag{8.28}$$

Values $\bar{Q}_j$ are installed according to the published data of the equipment or according to the design assignment; variance s^2_j of the weight of an individual piece of equipment can be ignored in most cases. Parameters $\bar{Q}, s_Q^2$ characterize the intensity of the load in the assembly of loaded individual cells of the production site. Average load on the area of a single cell is

$$\tilde{q}_\Delta = \frac{\tilde{Q}}{\bar{\Delta}}; \bar{q}_\Delta = \frac{\bar{Q}}{\bar{\Delta}}; s^2_{q_\Delta} = \frac{s_Q^2}{\bar{\Delta}^2} \tag{8.29}$$

The resulting load from $\tilde{n}_{QA}$ elements of technological equipment in the zone of influence A_φ will be equal,

$$\tilde{z} = \tilde{n}_{QA}\tilde{Q}; \bar{z} = \bar{\alpha}_Q n_A \bar{Q}; s_z^2 = \bar{\alpha}_Q n_A \left[s_Q^2 + \left(1 + \bar{\alpha}_Q\right)\bar{Q}^2 \right] \tag{8.30}$$

whence parameters $\bar{z}, s_z^2$ characterize the technological load in the assembly for all zones of influence of the production site; when used instead of $\bar{\alpha}_Q$ the value ω -quintile α^*_{QA} parameters $\bar{z}, s_z^2$ apply only to (*1*- ω)- the part of this assembly.

The process of loading, that is averaged over the area of the zone of influence, reads:

$$\tilde{q} = \frac{\tilde{z}}{A_\varphi} = \tilde{\alpha}_{QA}\tilde{q}_\Delta;\ \bar{q}_A = \bar{\alpha}_Q \bar{q}_\Delta;\ s^2_{q_A} = \frac{\bar{\alpha}_Q}{n_A}\left[s^2_{q_A} + \left(1 - \bar{\alpha}_Q\right)\bar{q}_\Delta^2 \right] \tag{8.31}$$

A randomly placed process of loads has a different effect on the internal forces and displacements of the structural element, depending on their type and the position of the sections to which they relate. This influence can be taken into account:

- by selecting a possible unfavorable position of the loaded area $\bar{\alpha}_Q A_\varphi$ (or $\alpha^*_{QA} A_\varphi$) in the area of determining the surface of influence $\varphi_z(x, y)$ considered effect *z*;
- also bringing actual loads with a known distribution density of random coordinates *f*(*x*,*y*) to uniform distributed loads equivalent in effect.

$$\begin{gathered} \tilde{z}_э = \tilde{z}_\phi \\ \tilde{z}_э = \tilde{q}_э \iint_{A_\varphi} \varphi_z(x, y) dxdy = \tilde{q}\bar{\varphi}_z A\varphi = \tilde{q}_э V_z \end{gathered} \tag{8.32}$$

where V_z volume limited by the surface of influence;

$\varphi_z = \frac{V_z}{A_\varphi}$ average value of the influence function.

The actual load effect of a randomly placed element with a resulting *Q*:

$$\tilde{z}_{\phi Q} = \tilde{Q}\varphi_z(x, y) \tag{8.33}$$

The equivalent load from condition (8.32) will be

$$\tilde{q}_{эQ} = \frac{\tilde{Q}\varphi_z(\tilde{x},\tilde{y})}{\varphi_z A_\varphi} \tag{8.34}$$

For a uniform density of random coordinates $f(x,y) = \frac{1}{A_\varphi}$

$$\bar{q}_{эQ} = \frac{\bar{Q}}{A_\varphi} = \frac{\bar{q}_\Delta}{n_A};\ s^2_{q_{эQ}} = \frac{\left(s^2_{q\Delta} + q^2_\Delta\right)\overline{\varphi^2_z}}{n^2_A\bar{\varphi}^2_z} - \frac{\bar{q}^2_\Delta}{n^2_\Delta};\ \overline{\varphi^2_z} = \frac{1}{A_\varphi}\iint_{A_\varphi}\varphi^2_z(x,y)\,dxdy \tag{8.35}$$

The actual load effect on behalf $\tilde{n}_{QA}$ elements of production equipment in the zone of influence A_φ occur:

$$\tilde{z}_\phi = \tilde{n}_{QA}\tilde{z}_{\phi Q} \tag{8.36}$$

The equivalent load is determined by the ratio (8.34):

$$\tilde{q}_э = \frac{\tilde{z}_\phi}{\bar{\varphi}_z A_\varphi} \tag{8.37}$$

The equivalent load parameters will be

$$\bar{q}_э = \bar{\alpha}_Q q_\Delta = \bar{q}_A$$
$$s^2_{qэ} = \frac{\bar{\alpha}_Q}{n_A}\left[\left(s^2_{q_\Delta} + \bar{q}^2_\Delta\right)\frac{\overline{\varphi^2_z}}{\bar{\varphi}^2_z} - \bar{\alpha}_Q\bar{q}^2_\Delta\right] \tag{8.38}$$

Coefficient of variation of the equivalent load becomes:

$$v_{qэ} = \frac{1}{\sqrt{n_A}}\sqrt{\left[\frac{1}{\alpha_Q}\left(1+v^2_Q\right)\frac{\overline{\varphi^2_z}}{\bar{\varphi}^2_z} - 1\right]} \tag{8.39}$$

where $v_Q = \frac{S_Q}{Q}$ determined according to Equation (8.28).

Coefficient of variation $v_{qэ}$ considers all the main factors of the technologic equivalent load: variability of load intensity from equipment elements, density of equipment distribution in the zone of influence, type of the considered effect in connection with the random location of the equipment of the zone of influence through $n_A = \frac{A_\varphi}{\overline{\Delta}}$.

8.6 COMBINATION OF LOADS AS RANDOM VALUES

This approach has been discussed in (Rzhanitsyn, 1978; Raizer, 1986). The design value of a single load is

$$F_{di} = \gamma_{F_i} F_r = \overline{F_i} \pm \beta_F s_{F_i} \tag{8.40}$$

where γ_{F_i} is the partial factor for actions, F_r is the representative load value, and $\overline{F_i}$ is the mean value of load, β_F – number of standard deviations. The negative sign in (8.40) is taken for unsafe situation when disuse takes place. The design force depends on the load linearly, then:

$$N_{di} = a_i F_{di} = a_i \overline{F_i} \pm \beta_F a_i s_{F_i} = \overline{N_i} \pm \beta_F s_{N_i} = \gamma_{F_i} N_{ri} \tag{8.41}$$

where N_i, s_{di} are the mean and standard deviation of the force.

$$s_{F_i} = \left| \frac{N_{ri}\left(\gamma_{F_i} - 1\right)}{\beta_F} \right| \tag{8.42}$$

For the design total force of several loads, factor β_N is taken the same as β_F for each single design load, that is, $\beta_N = \beta_F$, then:

$$N_d = \overline{N} + \beta_F s_N \tag{8.43}$$

If all loads are uncorrelated random quantities, then in accordance with the theorem on deviation of sum of random values (Gnedenko, 1969):

$$s_N = \left(\sum_{i=1}^{m} s_{N_i}^2 \right)^{1/2} \tag{8.44}$$

and for the total representative force according to the theorem about mean of sum of random values:

$$\overline{N_r} = \sum_{i=1}^{m} \overline{N_{ri}} \tag{8.45}$$

where m is the number of loads. Substituting (8.42), (8.44) and (8.45) in (8.43) yields:

$$N_d = \sum_{i=1}^{m} N_{ri} \pm \sqrt{\sum_{i=1}^{m} N_{ri} \left(\gamma_{F_i} - 1\right)^2} \tag{8.46}$$

Traditionally, the total force is considered as the sum of the separate forces:

$$N_d = \sum_{i=1}^{m} N_{di} = \sum_{i=1}^{m} N_{ri} \pm \sum_{i=1}^{m} N_{ri} \left(\gamma_{F_i} - 1\right) \tag{8.47}$$

Comparing (8.46) and (8.47), one can see that the use of Equation (8.47) in design practice would result in additional structural materials. It follows from the fact that the arithmetic negative mean value (8.47) is usually larger than the mean value under the square root in (8.46) when m>2, whereas N_d in (8.46) is 10–15% less than N_d in (8.47). It also depends on the number of loads, m, in combination and with the share of each load.

The probabilistic dependence of two forces, for example the bending moment M and the axial force N in the column, can be approximated as direct regressions M by N or N by M. (*Regression is a causal model of the statistical relationship between two quantitative variables x and y. The regression model is based on the assumption that the value of x is a controlled value, the values of which are set during the experiment, and the value of y is observed too during the experiment.*) Using the regression M by N, one expresses the mathematical expectation of the moment corresponding to the design value of the axial force:

$$\bar{M}_p = M_H + r_{MN} s_M \frac{N_p - N_H}{s_H} \tag{8.48}$$

where r_{MN} is the correlation coefficient of the total force.

From (8.48), in view of Equations (8.47) and (8.42), one obtains

$$\bar{M}_p = \sum_{i=1}^{m} M_{Hi} \pm r_{MN} \sqrt{\sum_{i-1}^{m} M_{Hi}^2 (\gamma_{f_i} - 1)^2} \tag{8.49}$$

Similarly, using the regression N on M, one can express the expected average value of the normal force $\bar{N}_p$ corresponding to the calculated value of the moment M_p.

To determine the correlation coefficient, the expression of the correlation moment is written:

$$k_{MN} = r_{MN} s_M s_N \tag{8.50}$$

When multiple independent loads are applied:

$$k_{MN} = \sum_{i=1}^{m} k_{M_i N_i} = \sum_{i-1}^{m} r_{M_i N_i} s_{M_i} s_{N_i} \tag{8.51}$$

Considering that the forces caused by a single load are related by functional dependence, that is, from (8.50) and (8.51) using (8.42) and (8.44) it is possible to express:

$$r_{MN} = \frac{\sum_{i=1}^{m} M_{Hi} N_{Hi} (\gamma_{f_i} - 1)^2}{\sqrt{\left[\sum_{i=1}^{m} M_{Hi}^2 (\gamma_{f_i} - 1)^2\right]\left[\sum_{i=1}^{m} N_{Hi}^2 (\gamma_{f_i} - 1)^2\right]}} \tag{8.52}$$

Substituting (8.52) in (8.49) and replacing for generality M by N by X and Y, the following equation is obtained:

$$\bar{Y}_p = \frac{\sum_{i=1}^{m} Y_{Hi} \pm \sum_{i=1}^{m} X_{Hi} Y_{Hi} (\gamma_{f_i} - 1)^2}{\sqrt{\sum_{i=1}^{m} X_{Hi}^2 (\gamma_{f_i} - 1)^2}} \tag{8.53}$$

Thus, for each load combination, it is necessary to check two variants of the designed forces $\left(\bar{X}_p, Y_p\right)$ and $\left(X_p, \bar{Y}_p\right)$ also.

Forces $\left(X_p, \bar{Y}_p\right)$ are obtained by the formulas (7.47) and (7.53), and $\left(X_p, \bar{Y}_p\right)$; for the resulting ones, replace X with Y and vice versa.

8.7 COMBINATION OF EXTREME VALUES OF LOADS

Based on the assumption used in (Wen, 1977; Raizer, 1986), pulse peaks are considered to be random sequences appeared at random time intervals τ and random duration Δ. Variations of loads in time present the simple Poisson flow of events. The instants of loads occurrence are independent. The return periods and duration time for loads are independent random values. The forces depend on loads linearly.

$$N(t) = \sum_{i=1}^{m} N_i(t) = \sum_{i=1}^{m} a_i F_i(t) \tag{8.54}$$

Distribution of the maximum force N can be expressed as follows:

$$P_s\left(N_{\max}\right) = \sum_{m,n=1}^{+\infty} P_{s,n}\left(N_n \mid t\right) P_n(t) \tag{8.55}$$

where $P_n(t)$ is the probability of load occurrence in the time range of $|0,t|$; $P_{sn}(t)$ is the distribution function of N_i in one test. The applied load has a Poisson distribution:

$$P_{s,n}(t) = \frac{(vt)^n \exp(-vt)}{n!} \tag{8.56}$$

where $\overline{v} = v$ is the mean frequency of the load occurrence. For identical N_n, distribution is

$$P_{sn} N_n(t) = P_s^n(N) \tag{8.57}$$

Substituting (8.56) and (8.57) into (8.55) one can get:

$$P_s(N_{\max}) = \exp\{-vt[1 - P_s(N)]\} = \exp[-vtP_f(N)] \tag{8.58}$$

where $1\text{-}P_s(N) = P_f(N)$.

For several independent loads, the distribution function of the maximum force is:

$$P_s(N_{\max}) = \exp\left[-\sum_{i=1}^{m} v_i t P_f(N_i)\right] \tag{8.59}$$

If the loads were applied simultaneously and correspond to a Poisson distribution, then:

$$P_s(N_{\max}) = \exp\left[-\sum_{i=1}^{m} v_i t P_f(N_i) - \sum_{i=}^{m}\sum_{=j}^{m} v_{ij} t P(N_{ij}) - \sum_{i=}^{m}\sum_{=j=}^{m}\sum_{=k=}^{m} \cdots \sum_{l}^{m} v_{i,j,k,\ldots l} t P(N_{i,j,k,\ldots l})\right] \tag{8.60}$$

where v_{ij} is the frequency of two loads applied simultaneously. This frequency can be expressed through the mean frequencies v_1, v_2 and mean durations $\Delta_1 = \overline{\Delta_1}$, $\Delta_2 = \overline{\Delta_2}$ of each load, then:

$$v_{1,2} = v_1 v_2 (\Delta_1 + \Delta_2) \tag{8.23}$$

For m loads we get:

$$v_{i,j,k\ldots l} = v_i v_j v_k \ldots . v_l (\Delta_i + \Delta_j + \Delta_k + \ldots + \Delta_l) \tag{8.61}$$

Equation (8.60) can be transformed to

$$P_s\left(N_{\max}\right)=\exp\left[-\sum_{i=1}^{m} v_i tP_f\left(N_i\right)-\sum_{i=}^{m}\sum_{=j=1}^{m} v_i v_j\left(\Delta_i+\Delta_j\right)tP_f\left(N_{ij}\right)-\right.$$
$$\left.-\sum_{i=}^{m}\sum_{=j=}^{m}\sum_{=k=}^{m}\cdots\cdots\sum_{=l=1}^{m} v_i v_j v_k \ldots v_l\left(\Delta_i+\Delta_j+\Delta_k+\ldots+\Delta_l\right)tP_f\left(N_{i,j,k,..l}\right)\right] \tag{8.62}$$

where N_{ij} is the total force of two loads occurring simultaneously. It follows that the distribution function of the maximum force depends on each load in the total force, on the frequency and duration of each load, and on service life. If random values N_i, N_{ij} ... N_l have identical distributions, and also if $v_i = v_j \ldots = v_p$; $\Delta_i = \Delta_j = \ldots = \Delta_l$, then from (8.62) follows (Raizer, 1995):

$$P_s\left(N_{\max}\right)=\exp\left[-t\sum_{i=1}^{m} C_n^k v^k \Delta^{k-1} P_f\left(N_k\right)\right] \tag{8.63}$$

where $C_n^k = C(k,n)$ is the number of combinations of n by k.

8.8 COMBINATION OF LOADS IN THE FORM OF MARKOV PROCESS

The assumption that the process of loading is the Markov process essentially expands the classification of loads that can be used for the structural analysis (Raizer, 1986). A process is Markov's by the intensiveness and time instants if the loads are consecutive and the instants of their occurrence depend on previous actions only and do not depend on the preceding load history. The discrete Markov chain represents the Markov random process, the space of whose states is finite (or countable). The multitude of indexes $T(1,2 \ldots)$ which is run in time coincides with the multitude of the integers or with part of it.

The analysis of the Markov chain is associated with the probability of its possible realizations. Important characteristics of these realizations are the transition probability matrix with k steps $P^{(k)} = \left\|P_{ij}^{(k)}\right\|$ where $P_{ij}^{(k)}$ is the probability of the transition to be realized through k steps from state i to j. When the one-step transitional probabilities are independent of a time variable (i.e. of value of k), the process possesses the stationary transitional probabilities, and then P_{if} is the probability of transiting from state i to state j during the testing period. Let us consider time-dependent loads $F_1(t)$ and $F_2(t)$, which can be presented as the Markov process with a countable number of states. The processes are assumed to be homogeneous. Shown in Figure 7.8 are states of first load and second load, respectively, $F_{ij} = i\Delta F_1 (i = 0,1,2,\ldots)$, $F_{2j} = j\Delta F_2 (j = 0,1,2,\ldots)$.

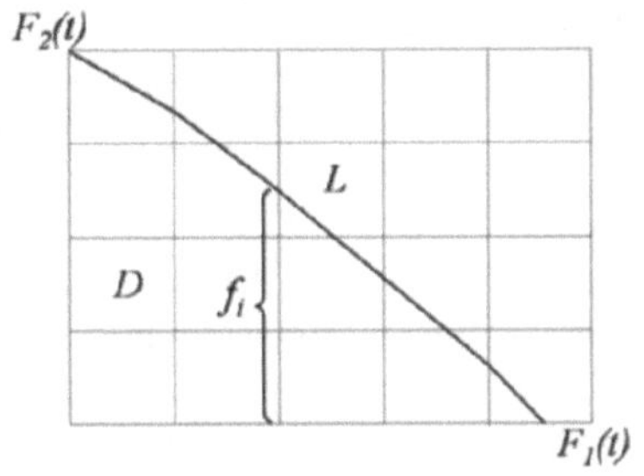

FIGURE 8.8 Possible states of two loads

Random processes are considered independent. The points in the plane in Figure 8.8 with coordinates (F_i, F_j) characterize possible states of some Markov homogeneous process. $F_1(t)$ and $F_2(t)$ can be imagined as simple birth and death processes (Gnedenko et al., 1969). Probabilities of states $P_i(t)$ of process $F_1(t)$ satisfy the Kolmogorov equations

$$\left.\begin{aligned} \frac{dP_0}{dt} &= -\lambda_0 P_0(t) + \mu_1 P_1(t) \\ \frac{dP_1}{dt} &= \lambda_0 P_0(t) - (\lambda_1 + \mu_1) P_1(t) + \mu_2 P_2(t) \\ \frac{dP_k}{dt} &= \lambda_{k-1} P_{k-1}(t) - (\lambda_k + \mu_k) P_k(t) + \mu_{k+1} P_{k+1}(t) \end{aligned}\right\} \quad (8.64)$$

Intensiveness of the transition probabilities λ_1, λ_2, λ_3 ... and μ_1, μ_2, μ_3 ... *is* considered to be *known*. The Kolmogorov equations for probabilities $G_0(t)$, $G_1(t)$, $G_2(t)$, of the states for $F_2(t)$ process can be written in the same way as (8.64).

$$\left.\begin{aligned} \frac{dG_0}{dt} &= -\rho_0 G_0(t) + \psi_1 G_1(t) \\ \frac{dG_1}{dt} &= \rho G_0(t) - (\rho_1 + \psi_1) G_1(t) + G_2(t) \\ \frac{dG_k}{dt} &= \rho_{k-1} G_{k-1}(t) - (\rho_k + \psi_k) G_1(t) + \psi_{k+1} G_{k+1}(t) \end{aligned}\right\} \quad (8.65)$$

The two-dimensional Markov process [$F_1(t)$, $F_2(t)$] is defined by the probability of state $\theta_{ij}(t)$ showing that the process is in (F_{1i}, F_{2j}). The Kolmogorov's equation for the two-dimensional process reads:

$$\begin{aligned} \frac{d\theta_{ij}}{dt} &= (\lambda_i + \mu_i + \rho_j + \psi_j)\theta_{ij}(t) - \lambda_{i-1}\theta_{i-1,j}(t) - \mu_{i+1}\theta_{i+1,j}(t) \\ &\quad - \rho_{i-1}\theta_{i,j-1}(t-1) - \psi_{i+1}\theta_{i,j+1}(t) \end{aligned} \quad (8.66)$$

Initial distributions are: if $t = 0$, then

$$\theta_{00}(t) = 1; \text{ if } i^2 + j^2 > 0, \text{ then } \theta_{ij}(t) = 0 \tag{8.67}$$

This distribution corresponds to the situation when at instant $t = 0$ there are no loads. Region D with boundary L (Figure 8.8) is considered here. This region can constitute, for example, the area of permissible state of cross-section (of a beam) with bending moment due to load $F_2(t)$ plotted to the y-axis versus normal force due to load $F_1(t)$ plotted at the x-axis. It should also be noted that all the pairs of values (F_{1i}, F_{2j}) which belong to region D are permissible and all pairs that don't belong to D are not permissible.

For an auxiliary two-dimensional random process with parameters (λ^*_i, μ^*_i) and (ρ^*_j, ψ^*_j), it is obvious that $\lambda^*_i = \lambda_i$, $\mu^*_i = \mu_i$ if in addition to point (F_{1i}, F_{2j}) region D includes also point ($F_{1,i+1}$, F_{2j}), or $\lambda^*_{i+1} = \mu^*_{i+1} = 0$ otherwise. Similarly, $\rho^*_j = \rho_j, \psi^*_j = \psi_j$,

if in addition to point

$$(F_{1j}, F_{2j})$$

region D, includes also point

$$(F_{1j}, F_{2j+1})$$

or $\rho^*_{j+1} = \psi^*_{j+1} = 0$ otherwise.

Before exiting from region D, the auxiliary process behaves in the same way as the main process. But after crossing the boundary of permissible states region, the process will remain beyond it. Probabilities of permissible states will satisfy the system of differential equations.

$$\frac{d\theta_{ij}(t)}{dt} = \left(\lambda^*_i + \mu^*_i + \rho^*_j + \psi^*_j\right)\theta_{ij}(t) - \lambda^*_{i-1}\theta_{i-1,j}(t) - \rho^*_{j-1}\theta_{i,j-1}(t) - \psi^*_{j+1}\theta_{i,j+1}(t) \tag{8.68}$$

where i and j represent all points in region D. The sum of probabilities of permissible states $\theta_{ij}(t)$ is the probability of no failure during the time t. System (8.68) must be solved together with the initial conditions (8.67). As an example, this solution was developed for a combination of snow and wind loads applied to a frame structure under snow, wind, and crane loads.

The snow load is represented as a Poisson process. The intervals of snow cover height 0 = 5, 5–15, 15–25 cm, and so on are taken as possible states of the process. As a result of statistical processing, the intensity of the transition probability ρ turned out to be 0.071/day. The transition from the height of the snow cover to the weight of the snow was made by choosing the snow density and the duration of the winter period T.

As a model of the random process of wind speed, it is proposed to use the birth and death process. Intervals were selected for possible process states 0–0,5, 0,5–1,5,

1,5–2,5 m / sec., and so on. An attempt was made to use a process model with two independent parameters λ and μ. By numerical analysis, the values $\lambda = 19.61$/day and $\mu = 301$/day were selected.

For the permissible state numbered i of the snow load in the range of permissible states is n_i acceptable values of wind speeds. The number of possible states of a two-dimensional process is $N = n_0 + n_1 + \ldots + n_m$. The Kolmogorov equations with respect to N unknowns is represented as

$$\frac{d\theta(t)}{dt} = A\theta(t) \tag{8.69}$$

Where A is a matrix of order N, defined by the structure of the equations (8.68); $\theta(t)$- a column vector.

$$\theta(t) = \left\| \theta_{00}(t), \theta_{01}(t), \ldots, \theta_{10}(t), \theta_{11}(t), \ldots, \theta_{m,n-1}(t) \right\| \tag{8.70}$$

In Raizer (1986) an algorithm contains for solving equations (8.69) that takes into account the features of the matrix A. So, when

$$m = 2, n_0 = 7, n_1 = 6, n_2 = 5$$

probability of intersection with the boundary L of the range of acceptable values

$$\theta(t) = 1 - \sum_{i=0}^{2} \sum_{j=0}^{n_i - 1} \theta_{ij}(t) \tag{8.71}$$

at t = 1,2,3,4,5 days, it turned out to be equal $\theta(t)$ = 0,20202; 0,40741; 0,56172; 0,67615; 0,76074.

In all solutions discussed above, only a linear relation between forces and loads was taken. There are some methodical difficulties in solving nonlinear problems. Therefore, the theory of plasticity can be applied provided that variations of all loads should be proportional to a single parameter. The probabilistic limit state design method was successfully developed to avoid these requirements.

9 Properties of Materials and Structural Deterioration

"I think you're begging the question," said Haydock, and I can see looming ahead one of those terrible exercises in probability where six men have white hats and six men have black hats and you have to work it out by mathematics how likely it is that the hats will get mixed up and in what proportion. If you start thinking about things like that, you would go round the bend. Let me assure you of that!

Agatha Christie (1890–1976), *The Mirror Crack'd*

9.1 GENERAL COMMENTS

Properties of materials and their statistical variations should be obtained from testing appropriate specimens. If the strength of a material can be assessed prior to using it in a structure, the characteristic strength (a specified fractal of the strength's statistical distribution produced according to the relevant standard) can be evaluated based on adequate statistical processing of available results. When this is not possible, for example, because the material will be produced on site of for any other reason, obtaining the specified characteristic strength should be ensured by adequate production control and acceptance procedures. In general, some information on the standard deviation or coefficient of variation is usually available and may be specified in codes of practice.

If the testing is carried out on a sufficient number of specimens, the value of resistance should be evaluated statistically. The characteristic value of material property corresponds in general to the lower fractal value in the material property distribution in the structure. The characteristic value of material properties as defined in (NKB, 1978) is the one that has an a priori accepted probability of not being reached at a hypothetically infinite test series (corresponding to a fractal in the distribution of the resistance parameter). This corresponds to a lower characteristic value. From the reliability point of view, the control of characteristic values should ensure that the strength of materials in the structure is, with a reasonable probability, in accordance with the assumptions used for the characteristic values in the design. Depending on the type of material, the quality control rules should distinguish between production control and compliance or identification control.

DOI: 10.1201/9781003265993-9

In order to ensure that the quality should not be too low, the quality control should be elaborated such that lots with characteristic value lesser than the required value has a low acceptance probability. The quality control should thus be such that materials, which on an average have a quality corresponding to the characteristic value, are likely to be accepted, so that the producer would be motivated to produce a quality with mean value complying with the required characteristic value. The observance of these two principles implies that the quality control should not only be based on a small number of tests but also take into account prior information concerning the standard deviation or coefficient of variation of material properties. It should be noted that if the materials are delivered from the production site with acceptable production control, a small-sample test ensuring against gross errors might treat the in situ control. When determining the values of materials properties, environmental conditions, which may cause deviations from the values obtained by testing, should be considered as well. The environmental conditions may include, for example, temperature and humidity. Effects of long-term actions should also be considered. If the design is based on resistance tests of structural members or calculations of values of material properties, which cannot be realistically determined, special statistical criteria should be applied on a case-by-case basis.

No less important for many industries is the problem of structural durability.

Assessing the reliability of a structure deteriorating in time is a very important and challenging problem in design. The deterioration of structural material results in decreasing the load-bearing capacity in time and thus increasing the probability of failure. Irreversible change of structural material properties can be caused by corrosion in steels, decomposition in wood, ageing in polymers, as well as abrasion or erosion and accumulation of defects. Though the physics and mechanics of the material deterioration processes are different, their effects on material processes and structural reliability have some similarities that make it possible to apply more or less similar approaches. Therefore, the models and peculiarities of corrosion wear and its effects on reliability are discussed in this chapter.

9.2 BAYESIAN TREATMENT IN ANALYSIS OF MECHANICAL PROPERTIES

The problem of reliability evaluation of manufactured structures or specimens is widely discussed with regard to controlling test results based on Bayesian treatment (Uritckiy, 1973; Raizer, 1986). For a substandard structural material supplied to a customer, we assume that the product belongs to a sample population. The controlled parameter is considered to be a random variable x with the given distribution function. For discrete distributions, when the number of sample population is not large, there is a $P(A_i/x_1, x_2, \ldots x_n)$ that the obtained results belong to the ith sample population:

$$P(A_i/x_1,x_2,\ldots,x_n) = \frac{P(A)P(x_1,x_2,\ldots,x_n/A_i)}{\sum_{j=1}^{k} P(A_i)P(x_1,x_2,\ldots,x_n/A_j)} \tag{9.1}$$

where $P(A_i)$ is the probability of the ith sample population to be submitted for control tests; $P(x_1, x_2, \dots x_n / A_i)$ is the probability that the control tests of the ith sample produce results $x_1, x_2, \dots x_n$; and k is the number of samples in the basic population. Changing to continuous densities, Equation (9.1) can be written as

$$p(y / x_1, x_2, \dots, x_n) = \frac{p(y)p(x_1, x_2, \dots, x_n / y)}{\int_{-\infty}^{\infty} p(y)p(x_1, x_2, \dots, x_n / y)\,dy} \tag{9.2}$$

where $p(y)$ and $p(y|x_1, x_2, \dots, x_n)$ are probability densities. Since the test results are independent, the joint conditional density is equal to the product of marginal conditional probabilities

$$p\left(x_1, x_2, \dots, x_3 / A_i\right) = \prod_{m=1}^{n} p(x_m / A_i) \tag{9.3}$$

Using (9.1), one can define the probability $P(x<C|x_1, x_2 \dots x_n)$ that the given sampling (e.g. of steel specimens) produced test results $x_1, x_2 \dots x_n$ which are less than the given value of C.

$$P(x < C / x_1, x_2, \dots, x_n) = \sum_{i=1}^{k} P(A_i / x_1, x_2, \dots, x_n) P(x_i < C) \tag{9.4}$$

where $P(x_i<C)$ is the probability that in the ith sample the test result was less than C.

Probability density in the basic population is marked as $p_y(x)$ while $p_i(x)$ is the probability density inside an arbitrary i^{th} sample population. For continuous densities, the expression can be rewritten as:

$$p(x < C / x_1, x_2, \dots, x_n) = \int_{-\infty}^{\infty} p(y / x_1, x_2, \dots, x_n) \int_{-\infty}^{C} p_i(x)\,dx\,dy \tag{9.5}$$

Using Equation (9.2) formula (9.4) reads:

$$P(x < C / x_1, x_2, \dots, x_n) = \frac{\int_{-\infty}^{\infty} p(y) \prod_{m=1}^{n} p(x_m / y) \int_{-\infty}^{C} p_i(x)\,dx\,dy}{\int_{-\infty}^{\infty} p(y) \prod_{m=1}^{n} p(x_m / y)\,dy} \tag{9.6}$$

Studies of control tests show that for basic and sample populations, it is possible to use the normal distribution for the yield stress and other characteristics of mechanical properties. If the variation range for deviations of sample population is not large, the deviation value can be taken constant and equal to $s^2(x_i)$. In this case:

$$p_i(x) = \frac{1}{\sqrt{2\pi s_{x_i}}} \exp\left[-\frac{(x-\bar{x}_i)^2}{2s_{x_i}^2} \right] \tag{9.7}$$

is the probability density function of controlled characteristic in sample population;

$x_1 = z$ is the mean value of controlled parameter in sample population considered to be a random variable with the density

$$p_g(x) = \frac{1}{\sqrt{2\pi s_{x_g}}} \exp\left[-\frac{(x-\bar{x}_g)^2}{2s_{x_g}^2} \right] \tag{9.8}$$

Equation (9.7) presents the density function of controlled characteristic with mean value $\bar{x}_c$ and deviation $s^2(x_1)$ in the basic population. It is proposed that the mean value $\bar{x}_i$ for controlled parameter in the sample population is a normally distributed random value z with the mean $\bar{z} = x_c$ and deviation according to the formula

$$s^2(z) = s^2(x_c) - s^2(x_1) \tag{9.9}$$

$$p(z) = \frac{1}{\sqrt{2\pi s_z}} \exp\left[-\frac{(z-\bar{z})^2}{2s_z^2} \right] \tag{9.10}$$

Formula (9.10) presents density function of mean values in sample populations. Using Equations (9.1), (9.4), (0.6) and (9.8), one can get the equation for the sample population

$$p(y / x_1, x_2, \ldots, x_n) = p(y / \bar{x}_m, n) = \frac{1}{\sqrt{2\pi s_y}} \exp\left[-\frac{(y-\bar{y})^2}{2\pi s_y^2} \right] \tag{9.11}$$

where $\bar{y} = \dfrac{\bar{x}_q s^2(x_i) + n\bar{x}_m s^2(z)}{s^2(x_i) + ns^2(z)}$; $s^2(y) = \dfrac{s^2(x_i) + s^2(z)}{s^2(x_i) + n^2 s^2(z)}$

Taking into account (9.7), (9.8), (9.10) and (9.11), Equation (8.6) reads

$$P(x < C / x_m, n) = \frac{1}{\pi} \int_{-\infty}^{\infty} \exp(-\frac{t^2}{2}) \int_{-\infty}^{\lambda} \exp(-u^2) du dt \tag{9.12}$$

Where $\lambda = \dfrac{C - ts(y)\sqrt{2} - \bar{y}}{\sqrt{2}s(x_i)}$

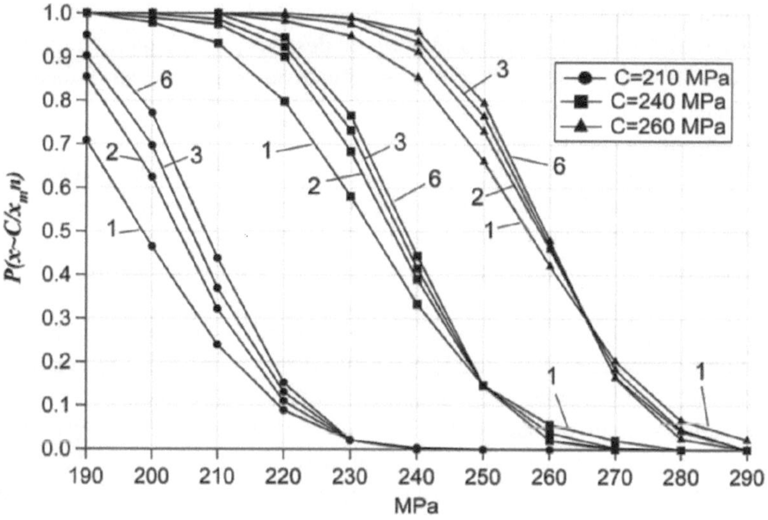

FIGURE 9.1 Relations for $P(x<C|x_{\mathrm{m}}, n)$

This is the main expression to calculate C in the controlled batch.

If $P(x<C|\overline{x}_{\mathrm{m}}, n) = 0.05$, then, for example, from (8.12), one can calculate the minimum value of $\overline{x}_m$, which will ensure that the condition $x \geq C$ will be satisfied with the probability of 0.95.

Let's consider an example for quality control of rolled steel by the mean value for sample of small volume. The values $\overline{x}$, $s(x_i)$, $s(z_i)$ vary within a rather wide range, close to experimental data for rolled sheet steel. Dependences

$P(x<C|\overline{x}_{\mathrm{m}}, n)$ on mean value of test results, sorting level C, number of tests n, parameters $s(x_i)$, $s(z)$ are presented in Figure 9.1.

Results of calculations allow making the following conclusion. If control tests results are equal to or exceed the sorting minimum C, then the probability of occurrence of substandard steel in this batch of rolling can be rather high. When C = 240 MPa and $\overline{x}_m$ = 250 MPa, this probability is equal to approximately 0.15 for different n.

9.3 SMOOTHING OF EXPERIMENTAL DATA

Experimental investigation of some random characteristic (yield stress, limit strains and so on) has been traditionally made by measuring limited numbers of samples. Results obtained in this case present sampling of random values from general aggregate. Empirical cumulative distribution function (CDF) and probability density function (PDF) expressed at the same time can't be used directly for describing the general aggregate. Experimental distribution function is a piecewise linear function, and its derivative, named the density function, will be a stepped function. In structural design problems all initial quantities are continuous. It follows, therefore, that for the probabilistic description of general aggregate, the empirical distribution function or density function should be smoothed, that is, approximated via some theoretical distributions.

Choosing an available theoretical distribution is a very complex problem. All the well-known goodness of fit does not confirm accuracy, taking into consideration the hypothesis about the type of distribution. Such a criterion can only turn down the considered hypothesis. Selecting one of many accepted hypotheses is always subjective. Sometimes it is possible to give a preference to a theoretical distribution law due to physical or technological conditions.

After selecting the distribution law, the problem is reduced to defining the distribution's parameters. This problem can be solved successfully by the method of moments, maximum likelihood method, least-squares method and others. If the distribution function is approximated as a whole, then the best solution can always be found in the region where high probabilities of random quantities appear, as the main information is concentrated exactly in such a region rather than in the distribution tails area.

The regular approximating method is unacceptable for solving reliability problems. The objective of reliability assessment is in calculating the failure probability

$$P_f = P_{rob}\left[R - F < 0\right] = \int_0^{\infty} P_R(x) p_F(x) dx \tag{9.13}$$

where $p_F(x)$ is the probability density of stresses (load effect). The integrand function in (9.13) essentially differs from zero at the cross-section of distribution tail areas of initial quantities. Approximating empirical distribution by a theoretical

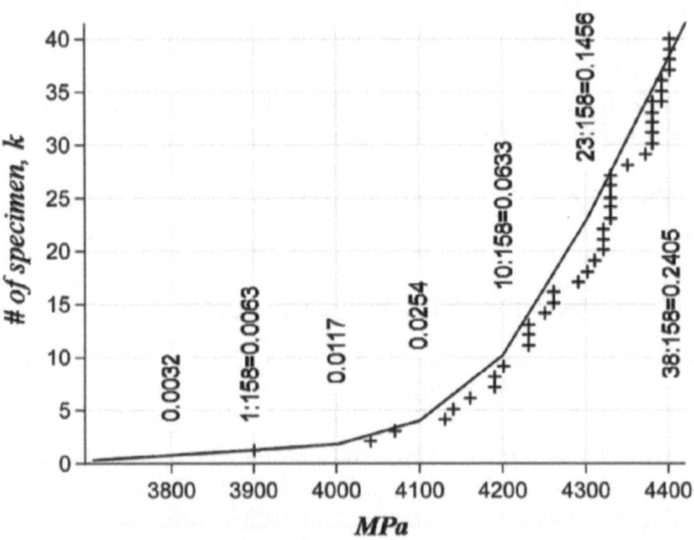

FIGURE 9.2 Empirical CDF of yield stress

one will be made in this section. It is not advisable to approximate all empirical distribution curves, when a structural reliability problem is considered. Local approximation should be made only in the distribution tail area. Application of the normal distribution can be recommended for a local approximation of empirical distribution functions. The local approximation with the use of the normal law is done in the following way (Rackwitz & Fiessler, 1978). Parameters of the approximating function should be found from the condition of equality of empirical (F_x^{ex}, f_x^{ex}) and approximating (F_x, f_x) distribution and their derivatives (densities) at some point, which can be called adjustment point x_a

$$F_x(x_a) = F_x^{ex}(x_a); \frac{dF_x(x)}{dx}\bigg|_{x=x_a} = \frac{dF_x^{ex}(x)}{dx}\bigg|_{x=x_a} \quad \text{or} \quad f_x(x_a) = f_x^{ex}(x_a) \tag{9.14}$$

Function $F_x(x) = \Phi(u)$ can be expressed through probabilistic integral

$$\Phi(u) = \frac{1}{\sqrt{2\pi}} \int_{-\infty}^{u} \exp(-t^2/2)dt, \quad u = \frac{x - \bar{x}}{s_x} \tag{9.15}$$

$$F_x(x_a) = F_x^{ex}(x_a) = \Phi(u_a)$$

Analogously,

$$u_a = \frac{x_a - \bar{x}}{s_x} \tag{9.16}$$

The same procedure for probability densities leads to:

$$\varphi(u) = \frac{1}{\sqrt{2\pi}} \exp(-u^{-2}/2) \tag{9.17}$$

$$f_x(x_a) = f_x^{ex}(x_a) = \frac{1}{s_x} \varphi(u_a) \tag{9.18}$$

From (9.16) we get:

$$u_a = \Phi^{-1}\left[F_x^{ex}(x_a)\right] \tag{9.19}$$

and from (9.17)

$$s_x = \varphi(u_a) / f_x^{ex}(x_a) \tag{9.20}$$

whereas (9.16) gives

$$\bar{x} = x_a - u_a s_x \tag{9.21}$$

Equations (9.20) and (9.21) define parameters of approximating distribution.

The CDF $P_{\sigma y}$ of yield stress σ_y was established as an envelope of 158 experimental points (Figure 9.2), which are the result of testing (Raizer, 2009) specimens of the steel used in the structure. The lowest test result was $\sigma_y = 390$ MPa, and the empirical CDF $P_{\sigma y}$ was assumed bounded on the left by the value $\sigma_y = 370$ MPa. Local approximation of this empirical probabilistic distribution function is presented in Figure 9.3.

Parameters of the approximating normal law can be defined by the method of moments. This approach allows us to consider mean value $\bar{\sigma}_y$, and standard deviation $s(\sigma_y)$ to be equal to the experimental values of σ_y^{ex} and $s^{ex}(\sigma_y)$. The result is: $\sigma_y = 468$ MPa, $s(\sigma_y) = 34$ MPa (dotted line in Figure 8.3). For local approximation two places of adjustment point are chosen:

at point $\sigma_y^a = 390$ MPa we get: $F_{\sigma_y}^{ex}(\sigma_y^a) = 0.0063$, $f_{\sigma_y}^{ex}(\sigma_y^a) = 0.000032$, $\Phi(u_a) = 0.0063$, $u_a = -2.495$, $\varphi(u_a) = 0.017749$, $\bar{\sigma}_y = 534$ MPa, $s(\sigma_y) = 55$ MPa.

These values are far away from the sample values. This means that in the areas of mean value and especially the "right" tail, this approximation is deemed too rough. But in the region important for probability of failure calculation the local approximation is essentially close to experimental data.

At adjustment point $\sigma_y^a = 405\ MPa$ we get: $F_{\sigma_y}^{ex}(\sigma_y^a) = 0.01855$, $f_{\sigma_y}^{ex}(\sigma_y^a) = 0.000137$, $\Phi(u_a) = 0.01855$, $u_a = -2.085$, $\varphi(u_a) = 0.045386$, $\bar{\sigma}_y = 474$ MPa, $s(\sigma_y) = 47$ MPa.

It is seen (Figure 9.3) that a good approximation for the different parts of empirical distribution function becomes possible due to different positions chosen for the adjustment point.

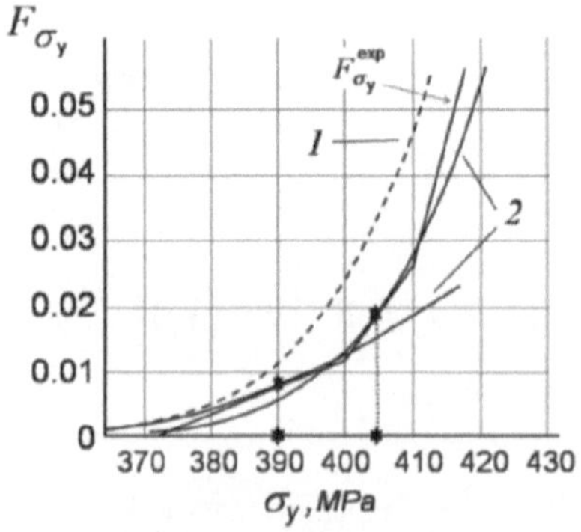

FIGURE 9.3 Local approximation of tail area distribution; 1—approximation by moments; 2—local approximation; *—adjustment point

9.4 RELIABILITY OF CORRODING STRUCTURES

A problem of structural durability and protection against local corrosion is acutely important and will be discussed in this section (Raizer, 1990). Local corrosion leads to local degradation visible on the metal surface in the form of rusty spots, ulcers, pits and cracks (Figure 9.4). The process of degradation is random. It can be described using the following assumptions:

- events that affect the occurrence of corrosion cavities at unrelated time intervals are independent;
- probability of corrosion cavity occurrence at an arbitrary time interval t is proportional to the extent of the time interval with proportionality factor μ;
- occurrence probability of two or more events within an extremely small time interval is an infinitely small value of a higher order of magnitude than that for either event.

A simultaneous realization of all these assumptions takes place for the simplest flow of events, also known as a uniform Poisson process. Such a process can be described by a system of differential equations:

$$\left.\begin{aligned} \frac{dP_0}{dt} &= \mu P_0 \\ \frac{dP_n}{dt} &= \mu(P_{n-1} - P_n) \end{aligned}\right\} \tag{9.22}$$

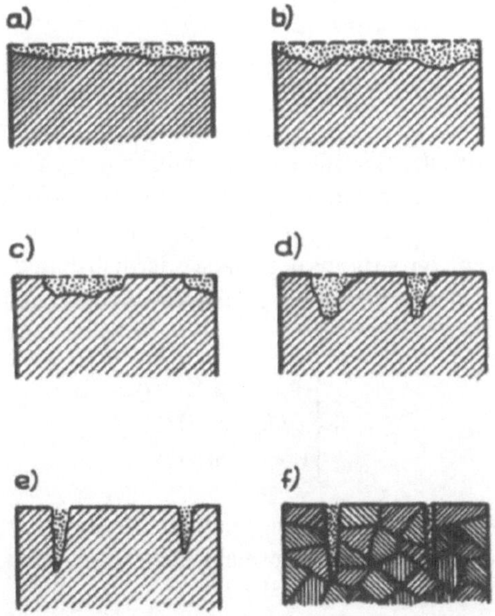

FIGURE 9.4 Types of structural steel corrosion: a) uniformly distributed; b) irregularly distributed; c) corrosion spots; d & e) pitted corrosion; f) corrosion cracks

The initial conditions for this system of equations are:

$$\left.\begin{array}{ll} P_n(t) = 1 & \text{when } \; n = 0 \\ P_n(t) = 0 & \text{when } \; n = 1, 2, \ldots \end{array}\right\} \tag{9.23}$$

There is only one solution for the system (9.22), and together with conditions (9.23) it can be presented as the Poisson distribution:

$$P_n(t) = \frac{[\mu(t - t_0)^n]}{n!\exp[-\mu(t - t_0)]} \tag{9.24}$$

Equation (9.24) is the Poisson distribution and constitutes the probability that in the instant $t \geq t_0$ the system is in the state of n ($n = 1, 2, 3, \ldots$). If the number of cavities occurring within a time interval can be simulated by the Poisson distribution, the time of the next cavity occurrence follows the exponential distribution (Kapur & Lamberson, 1977):

$$P(t) = \exp(-\mu t). \tag{9.25}$$

There is a rather limited volume of experimental data associated with studying the kinetics of pitting cavity formation and growth in structures, as well as on spread in their number. Experimental dependences were obtained by (Raizer,1990):

$$\mu = \mu_{gr}(1 - e^{-\beta t}) \tag{9.26}$$

where μ_{gr}, β are empirical coefficients. The value of μ_{gr}, measured as the number of defects per a unit of structural surface, varies within a wide range.

The most important parameters of irregular pitting corrosion include the maximum depth of a cavity, its diameter, and its area. The random value of cavity depth, δ_k (k – a random point on structural surface), is distributed in the final interval $[0, h_0]$, where h_0 is the structural element's thickness distributed uniformly, that is:

$$P_\delta(x) = \begin{cases} 0 & \text{at } \; x < 0 \\ x / h_0 & \text{at } \; 0 \leq x \leq h_0 \\ 1 & \text{at } \; x > h_0 \end{cases} \tag{9.27}$$

Distribution of the maximum depth for n cavities, that is, $\delta_n = \max\{x_1, x_2, x_3, \ldots, x_n\}$, is known from the theory of extreme values (Galambos, 1987) and can be taken as exponential:

$$P_{\delta_n} = \begin{cases} \exp[-n(h_0 - x)] & \text{at } \; 0 \leq x \leq h_0 \\ 1 & \text{at } \; x > h_0 \end{cases} \tag{9.28}$$

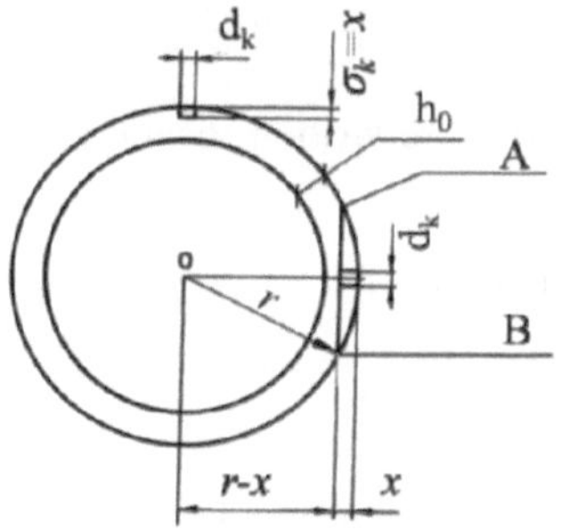

FIGURE 9.5 Cross-section of the pipeline with idealized cavity

The next important parameter is an across size of cavity or—assuming that it is of a circular shape—diameter (Figure 9.5). Let us assume the depth of cavity is equal to x. Then the possible range of diameter variation is the chord AB, whose length is $2\sqrt{2rx - x^2}$ and r is the external radius. An assumption is taken that the random value of diameter y_i has a uniform distribution within the interval $[0, 2\sqrt{2rx - x^2}]$:

$$P_d(y) = \begin{cases} 0 & \text{if } y < 0 \\ \dfrac{y}{2\sqrt{2rx - x^2}} & \text{if } y < 0 < 2\sqrt{2rx - x^2} \\ 1 & \text{if } y > 2\sqrt{2rx - x^2} \end{cases} \quad (9.29)$$

Distribution of the maximum diameter for n cavities $dn = \max(y_1, y_2, y_3, \ldots y_n)$ is:

$$P_{d_n} = \begin{cases} \exp[-n(2\sqrt{2rx - x^2} & 0 \le y \le 2\sqrt{2rx - x^2} \\ 1 & y > 2\sqrt{2rx - x^2} \end{cases} \quad (9.30)$$

The third parameter of cavity is its area A_k. Knowing the maximum area is important for solving the problem. There are some uncertainties about this in the theory of order statistics. The point is that the maximum value of δ_n doesn't necessarily correspond to the maximum value of d_n. Accepting this notion would lead to a more conservative solution. The third parameter of cavity is its area A_k. Knowing the maximum area is important for solving the problem. There are some uncertainties about this in the theory of order statistics The point is that the maximum value of δ_n doesn't necessarily correspond to the maximum value of d_n. Accepting this notion would lead to a more conservative solution. (i.e. the no-failure probability based on this notion would be less than the real one)—see case 1 below. If the notion is ignored, two kinds of versions can be offered: (a) distribution of maximum depth δ_n and, depending on this, distribution of diameter d_k for one cavity in the first version (see case 2 below), and otherwise (b) distribution of maximum diameter d_n and, depending on this, distribution of depth δ_k in the second version (see case 3 below).

Types of $PA(x)$ distributions are written for three cases:

$$\begin{aligned} &P_{\delta n}(x) = \exp[-n(h_0 - x)], \quad x \in [0, h_0], \\ &\text{Case 1}: P_{dn}(y) = \frac{y}{2\sqrt{2rx - x^2}}, \qquad 0 \le y \le 2\sqrt{2rx - x^2} \end{aligned} \tag{9.31}$$

The area of the cavity A_k equals the area of the segment in Figure 9.5:

$$A_k = r^2 \arcsin\frac{y}{2r} - \frac{y}{2}\sqrt{r^2 - \frac{y^2}{4}} + [x - r + \sqrt{r^2 - \frac{y^2}{4}}]y \tag{9.32}$$

The maximum possible value of the cavity area A_k is when $x = h_0$ and $y = 2\sqrt{2rh_0 - h_0^2}$. In large diameter pipes values of x/r and $y/2r$ are very small and the area can be approximated as $A_k = xy$; thus, it follows:

$$P_{A_k}(A) = \frac{An}{2\sqrt{2r}} \int_0^{h_0} \exp[-n(h_0 - x)]x^{-\frac{3}{2}} dx \tag{9.33}$$

where A_k is a random quantity uniformly distributed within [0,A*].

$$\begin{aligned} &P_{\delta_n}(x) = \frac{x}{h_0}, \qquad x \in [0, h_0] \\ &\text{Case 2}: P_{d_n}(y) = \exp\left[-n\left(2\sqrt{2rx - x^2} - y\right)\right], \quad y \in \left[0, 2\sqrt{2rx - x^2}\right] \end{aligned} \tag{9.34}$$

Distribution of A_k is:

$$P_{A_k}(A) = \frac{1}{h_0} \int_0^{h_0} \exp\left[-n\left(2\sqrt{2rx - x^2} - \frac{A}{x}\right)\right] dx \tag{9.35}$$

Case 3:

$$\begin{aligned} &P_{\delta_n}(x) = \exp\left[-n\left(h_0 - x\right)\right], \qquad x \in \left[0, h_0\right] \\ &P_{d_n}(y) = \exp\left[-n\left(2\sqrt{2rx - x^2} - y\right)\right], \quad y \in \left[0.2\sqrt{2rx - x^2}\right] \end{aligned} \tag{9.36}$$

$$\text{In follows}: P_{A_k}(A) = \int_0^{h_0} \exp\left(-2\sqrt{2rx - x^2} - \frac{A}{x}\right) d\left[\exp\left(-n\left(h_0 - x\right)\right)\right] \tag{9.37}$$

The last case, as mentioned above, yields a conservative solution.

Example 1. Reliability of a pipeline subjected to one-sided irregular corrosion.

Dimensions of the corrosion cavity depth and diameter gradually grow so that a failure of pipe will occur when a cavity would grow through the wall. Time t_n before this occurs can be calculated using the expression:

$$\int_0^{t_n} v(t)dt = h_0 - \delta_n \tag{9.38}$$

where δ_n is maximum depth among n cavities; $v(t) = v_0\exp(-\alpha t)$ is corrosion rate (Raizer, 2009). From (9.38) we get:

$$t_n = \frac{1}{\alpha}\ln\frac{v_0}{h_0 - \delta_n} \tag{9.39}$$

Time distribution $P(t_n < t)$ for a cavity to break through the wall can be expressed as:

$$P_n(t) = P\left\{\delta_n \geq \left[h_0 - \frac{v_0}{\alpha}\left(1-\exp\left(-\alpha t\right)\right)\right]\right\} = 1-\exp\left[-n\frac{v_0}{\alpha}\left(1-\exp\left(-\alpha t\right)\right)\right] \tag{9.40}$$

After averaging for n cavities, it follows:

$$P(t) = \sum_{n=0}^{\infty}\frac{(\mu t)^n}{n!}\exp\left\{1-\exp\left[1-\exp\left(-\alpha t\right)\right]\right\} \tag{9.41}$$

Example 2. Design of structural members under axial tension.

A cylindrical element having a ring cross-section is considered. This element is subjected to irregular corrosion under deterministic load F. Denoting A_0 as the initial value of cross-section ($t = 0$) and A_k as the area of cavity with given distribution $P_{Ak}(A)$ allow one to express the condition of no failure as follows:

$$\frac{F}{A_0 - A_k} < R_y \text{ or } A_k < A_0 - \frac{F}{R_y} \tag{9.42}$$

Substituting the last expression into a distribution function as an argument and averaging with respect to n and R_y the probability of no failure at instant t is obtained:

$$P(t) = \exp(-\mu t)\sum_{n=0}^{\infty}\frac{(\mu t)^n}{n!}\int_0^{\infty} P_{Ak}\left(A_0 - \frac{F}{R_y}\right)p\left(R_y\right)dR_y \tag{9.43}$$

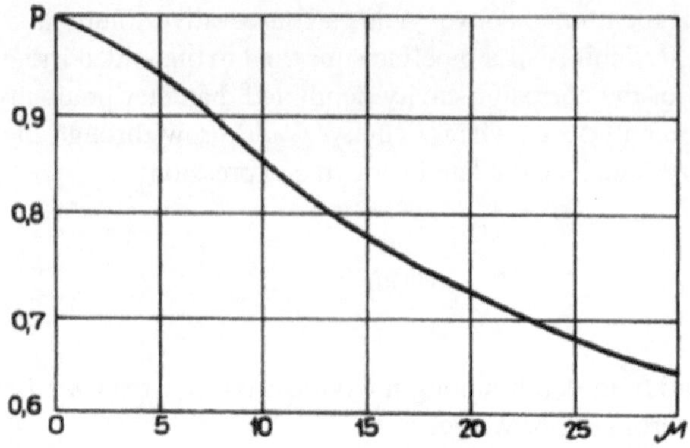

FIGURE 9.6 Reliability function for Example 1

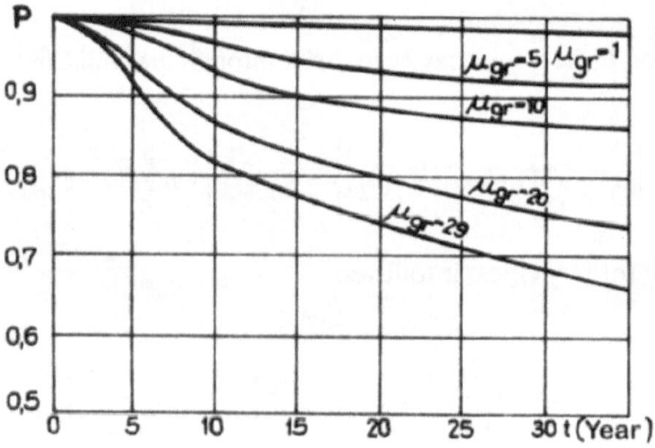

FIGURE 9.7 Reliability function for Example 2

where $p(Ry)$ is distribution density of yield stress. The following data are taken in the numerical example: external diameter $D = 6.26$ in; initial thickness $h_0 = 0.24$ in; $F = 58{,}08t$ $\mu = \mu_{gr}[1 - \exp(-\beta t)]$ and $\beta = 0.05$; $\bar{R}_y = 290$ Mpa; $s_{R_y} = 25$ Mpa. Parameters of cavity are $\bar{d}_k = \bar{\delta}_k = 0.008$ in. Results of numerical calculations are plotted in Figures 9.6–9.7.

10 Risk Evaluation and Optimal Probabilistic Design

Civil engineering is a high-risk business. Its professional institution in Britain was founded "to direct the great sources of power in Nature for the use and convenience of man". Neither Nature nor man are readily predictable, and it is therefore inevitable that the tasks fulfilling that aim are beset with hazard. Moreover, the design and construction of civil engineering works, without which the natural hazards of an uncivilized life would far exceed those experienced now, is undertaken within severe constraints.

(A.R. Flint, 1981)

Deterministic optimization has produced greatly improved performance in all types of engineering systems. It can, however, lead to unreliable design if the uncertainty is ignored. All systems exhibit uncertainty. This arises from (a) input uncertainty (e.g., loadings, supply voltages, etc.), and (b) component variation (e.g., resistance and dimensional tolerances, etc.) that results in performance responses) variations. Robust design is a methodology that attempts to ensure that the responses are insensitive to both the input uncertainty and component variations without actually eliminating the causes.

(Young Kap Son and Gordon J. Savage, 2007)

10.1 ACCEPTABLE RISK

The word *risk* is used in everyday life in two different albeit related meanings. We don't, as a rule, think about these distinctions in our ordinary conversations. What kind of concept presents the term risk? Let us consider two situations:

1. Two men are on their way to a meeting. One proposes to take a taxi. The other looks at his watch and says that they have ten minutes; if they walk fast, they will be on time, but a taxi might get stuck in a traffic jam and there is a risk of being late. The first argues that they would be in the taxi only for two minutes;

DOI: 10.1201/9781003265993-10

he doubts that they would spend eight minutes in a traffic jam, so the risk of being late is not so serious. Here the word *risk* is used in the sense of *probability of an undesirable event.*

2. Two businessmen are being offered a deal. One states: "A lucrative deal is proposed. The profit could be very high." The other counters: "No, I can't agree. It is too risky. If it fails, I will lose everything." Here the word *risk* means *loss through an undesirable event.*

Thus, the concept of risk contains two essential elements. Common to both is the uncertainty of an undesirable event or an unfavorable state.

In technical papers (Raizer, 2004) the term *risk* is used in both of these senses. In reality, no structure can be designed to be completely free from the risk of failure whatever the cost. If, however, the risk is taken into account, the losses can turn out to be insignificant in comparison with the cost. The following definition of risk appears to be most acceptable: *The value of risk is determined as a product of consequences due to the event times the measure of the event's occurrence possibility.* When the consequences of an event are negative, the value is a loss, and can usually be expressed in monetary terms. Thus the safety of the structure can be assessed quantitatively as:

$$R = P_f U \tag{10.1}$$

where P_f is the probability of failure; U stands for the losses incurred through it.

The same risk can be a combination of a high probability of failure and limited consequences, or of limited probability of failure and a high level of losses. It should be noted that as a rule, for some kinds of failure, its probability and consequences are inversely proportional. When multiple kinds of consequences are involved, they rarely fall within a single category. An influence function $Z(U)$, which reduces various consequences to a single (e.g. monetary) comparative basis (Spaethe, 1987), can be conveniently introduced:

$$R = \int Z(U) P_f dU \tag{10.2}$$

Construction activity in any industry should proceed with total compliance with the requirements of design regulations and technological procedures. It is commonly assumed that the structure would fail if these requirements are not satisfied. The probability of failure is the main numerical measure of the structure's reliability. For the latter, it would be irrational to set a threshold too low to be exceeded in all cases. One should bear in mind, however, that the higher the reliability level, the higher would be the construction cost. On the other hand, the lower the probability of failure, the lower would be the cost of repair and restoration. In other words, a rational solution can be found as the "golden mean," that is, an expedient level for the failure probability.

If the consequences of failure can be estimated in monetary terms alone, an expedient level can in principle be determined through minimization of the objective function

$$C = C_0 + \sum R \tag{10.3}$$

where C is the total cost of construction of structure and its maintenance during service life; C_0 is the initial investment; and R is the summary risk for all types of failures, that is, the mean total probable losses due to them.

$$\sum_i R_i = \sum_i P_{fi} U_i \xi(t) \tag{10.4}$$

where P_{fi} is the probability of the i-th type failure; Ui denotes the losses due to such a failure; $\xi(t)$ is a discounting function, that is, a function allowing the expenditures at different years to be reduced to that at the time of commissioning. It should be noted that not always the consequences of failure could be estimated in monetary terms alone. Examples on the contrary include ecological disasters, injuries and fatalities, damage to historical monuments, and many other events that may carry non-obvious potentials of political and social repercussions.

Regarding the monetary equivalent of the non-economic part of risk, it was proposed to define the level of cost that can be acceptable for human safety in the society at the given stage of its development. This value can then be used formally in the objective function as "the cost of human life." For a situation involving a single possible type of failure, the objective Equation (10.3) can be written as follows:

$$C = C_0 + P_f U \tag{10.5}$$

where U is as per Equation (10.4), and the discount function is irrelevant. The optimal reliability level will be characterized by the optimal failure probability P_f^{opt} which can be found from the condition:

$$\frac{dC}{dP_f} = 0 \tag{10.6}$$

With quantity U being independent of P_f, we have

$$\left.\frac{dC}{dP_f}\right|_{P_f = P_f^{opt}} = -U \tag{10.7}$$

For a structure its construction cost C_0, risk R and total (construction and maintenance) cost C depend on probability of failure as shown in Figure 10.1. The minimum cost is

$$C_{\min} = C(P_f^{opt}) \tag{10.8}$$

Assuming that besides the losses U as per equation (10.1), the consequences of failure include N casualties, and setting P_fN as the mean equivalent loss, we have

$$v(P_f) = (1 - P_f)N \tag{10.9}$$

If the design refers only to the economic optimal reliability level, the mean value of live protection will be

$$v(P_f^{opt}) = (1 - P_f^{opt})N \tag{10.10}$$

In this case the optimized function is associated with only part of the risk. If human life could have monetary expression and be included in the objective function, then the optimal failure probability should be low. The reason is that increasing the steepness of line e in Figure 10.1 would shift the minimum of function C to the left. It is naturally expected that increased reliability levels of structures increase overall human safety. Dividing the total cost C by the number of "protected lives," we obtain the per-capita cost as:

$$e = \frac{c}{v} = \frac{C_0 + P_f U}{\left(1 - P_f\right)N} \tag{10.11}$$

The condition for minimum of $e(P_f)$ is

$$\frac{de}{dP_f} = 0 \tag{10.12}$$

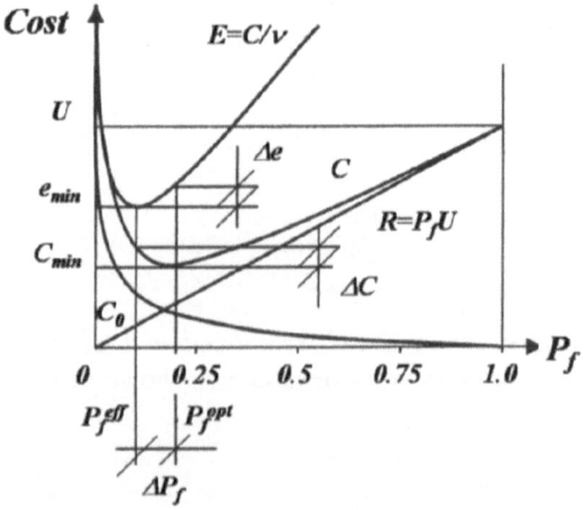

FIGURE 10.1 Analysis of effectiveness in increasing safety

which will make sense if

$$\frac{de}{dP_f}=\frac{\frac{dC_0}{P_f}\left(1-P_f\right)+C_0+U}{N\left(1-P_f\right)^2}=0 \tag{10.13}$$

The value of failure probability P_f with condition (10.13) satisfied can be called the *effective* value. It is given by

$$\frac{dC_0}{dP_f}_{P_f=P_f^{ef}}\left(1-P_f^{ef}\right)=-\left(C_0+U\right) \tag{10.14}$$

and is independent of size N of the exposed population.

If we ignore an optimal solution (Figure 10.1) and reduce the probability of failure by the value of $\Delta P_f = P_f^{opt} - P_f^{ef}$, the mean value of live protection will increase by $\Delta v = v(P_f^{ef}) - v(P_f^{opt})$, and the total cost of the structure will increase by $\Delta C = C(P_f^{ef})$. Thus, C_{min} and the live protection cost per capita will decrease by

$$\Delta e=\frac{C_{\min}}{\left(1-P_f^{opt}\right)N}-\frac{C\left(P_f^{ef}\right)}{\left(1-P_f^{ef}\right)N} \tag{10.15}$$

Further decrease of the probability of failure will increase the size of protected population, as well as the total cost of structure and the "per-capita' cost: in other words it will reduce efficiency of investment. Therefore, the neighborhood of points $P_f = P_f^{ef}$ on the x-axis in Figure 10.1 can be considered a domain of the maximum efficiency of economic response to non-economic risk. It would be reasonable to assume that the optimal design solution would fall within this domain, so that the interval $\Delta P_f = P_f^{opt} - P_f^{ef}$ restricts the area of designer's freedom.

As mentioned above, risk could be defined either as the probability of failure, or as a scale of possible losses through failure, or as a combination of both. Thus, its value can be considered to be a vector consisting of several components, that is why there is a multi-parametric choice. Intuitive feeling suggests that there is a risk level, which can be considered acceptably low after which there is no need to invest more for higher safety. At the other extreme, there is a level of the maximum acceptable risk, which cannot be reduced regardless of the cost. Between these two levels, there is a domain where a compromise between cost and non-economic considerations can be found. If greater safety is advisable, a solution can also be found in the form of an alternative construction method, failure prevention measures, provisions for evacuation, and so on.

10.2 OPTIMIZATION OF RELIABILITY INDEX

"OPTIMAL" means "BEST." It is derived from the Latin word "OPTIMUS." Thus, optimization can be regarded as aspiration for the best. In all practical applications one encounters the same problem: how to choose the best from many different options. This idea corresponds to Leonard Euler's (1707–1783) philosophy expressed as follows:

> Since the building of the universe is perfect and is created by the wisdom of the Creator, nothing arises in the universe in which one cannot see the sense of some maximum or minimum.

Development of the theory of reliability combined with engineering experience makes it possible to foresee the future steps in structural design based on the method of probabilistic optimization (Rzhanitsyn, 1978; Raizer, 1986). The optimal design procedure based on probabilistic description of construction and operation consists of exterminating the design parameters, which in turn should provide conditions of acceptable maintenance with economic efficiency during the structure's service life. This can be reached by minimizing the total cost of construction and maintenance of the structure during its service life. The general aim of any design consists of ensuring both reliability and economic efficiency. But these tusks are contradicting. A rise of reliability leads to a higher price of structure (economic efficiency decreases), and otherwise a reduction in price leads to a decrease in reliability. This problem has been usually solved intuitively by common-sense judgment. There is nothing wrong in using professional intuition since it is usually based on past experience. And as a step further, the design of structures with sufficient degree of reliability at justifiable cost is possible with the help of the optimal probabilistic method.

The most widespread reliability factor is the probability of no failure (or survival) during the service life of the design. In the general theory of reliability (Gnedenko et al., 1969) this factor represents the loss by structure of its service quality. Under the general principle of design at maximum economic efficiency (or at minimum cost of construction and maintenance) the term *loss of quality* should also be understood as *loss of cost*. In this case a failure constitutes a random event accompanied by some losses. It takes place when the structure is designed, manufactured, and maintained (or either of that) in violation of appropriate requirements. Some losses can be expressed in monetary terms, while others are difficult or impossible to express in such terms (like those associated with ecological catastrophes, casualties, etc.).

Different types of failures with corresponding losses can happen in the same structure. The following classification can be applied. If the losses can be expressed in monetary terms, this category can be called "structures with only economic responsibility." If the losses cannot be so expressed, the category can be called "structures with non-economic responsibility." There is also third type which can be called "structures with mixed responsibility."

The principle of maximum economic efficiency for the first category consists in minimizing the mean cost of the design service life. This mean cost represents the construction cost plus the mean of all losses due to failures during the service life. Components of this sum depend on the probability of no failure. When this probability increases the construction cost increases, while expenses of repair and restoration decrease, that is, an optimization problem arises. This optimal probabilistic problem can be solved by minimization of the target function:

$$C = C_0 + \sum_{j=1}^{m} \int_0^T U_j(t)\xi(t)dt \tag{10.16}$$

where C is the total cost of construction and compensation for losses due to possible damage and failures during the design service life; C_0 is the construction cost; m is the number of different types of failures; U_j is the mean rate of loss accumulation due to the j-type failure; $\xi(t)$ is a cost depreciation function, which reduces all the costs to the currency rate at the time in question; T is the design service life of structure.

Essential difficulties arise in probabilistic optimization of structures without economic responsibility in expressing the cost of non-economic losses. In principle, the answer can be found in aspiration for optimal use of the existing resource subject to maximum possible safety for all objects.

In any case determining the optimal level of structural reliability can be reduced to minimizing the target function of cost. The unknown optimal parameters of design can include the no-failure probability, design values of loads or resistance, area of cross-section, and so on. In the case of unexpected failure (e.g., loss of strength in a structural member) the target function can read:

$$C = C_0 + \sum_{j=1}^{m} U_j P_{if} = \min \tag{10.17}$$

where Uj is the loss due to each failure; P_{jf} is the probability of individual failure.

It is implied in (10.17) that if the probability of a failure is low, that of recurring failures can be assumed negligible. The minimization condition for Equation (10.17), given that failure probabilities are independent, reads:

$$\frac{\partial C_0}{\partial P_{if}} + \sum_{j=1}^{m} \left(U_j + P_{jf} \frac{\partial U_j}{\partial P_{jf}} \right) = 0 \tag{10.18}$$

Let us consider a type of failure when a rupture of a tension bar takes place. Losses can be expressed in the form:

$$U_j = \chi_{id} + \chi_d A \tag{10.19}$$

where χ_{id} is the indirect losses, independent of the cross-section dimensions, $\chi_d A$ is the direct losses due to restoration that depend on the cross-section area A, χ_d is a coefficient. The construction cost, C_0, is given by:

$$C_0 = C_0^* + \alpha A \tag{10.20}$$

Substituting (10.19) and (10.20) in (10.18), we have:

$$(\alpha + \chi_{id} P_{if})(dA / dP_{if}) + \chi_{id} + \chi_d A = 0 \tag{10.21}$$

The random reserve of strength $\tilde{S}$ is given by:

$$\tilde{S} = A\tilde{R} - \tilde{F} \tag{10.22}$$

where $\tilde{R}$ is the random limit of resistance, and $\tilde{F}$ is the random force due to applied load.

If R and F are normally distributed random values, the failure probability can be given by:

$$(P_f(S) = 0.5 - \Phi(\beta)\beta) \tag{10.24}$$

where $\Phi(\beta) = \frac{1}{\sqrt{2\pi}} \int_0^\beta \exp(-x^2/2)dx$ and the reliability index β are given by:

$$\beta = \frac{A\bar{R} - \bar{F}}{\sqrt{A^2 s^2(R) + s^2(F)}} \tag{10.25}$$

where $\bar{R}$ and $\bar{F}$ are the average values, and $s(R) = s_r$ and $s(F) = s_f$ are the standard deviations.

Then equation (10.21) can be rewritten in the form:

$$\alpha + \chi_d \left\{ \frac{1}{2} - \Phi\left[\frac{z-1}{\sqrt{(v_r^2 z^2 + v_f^2)}} \right] \right\}$$
$$= \Phi'\left[\frac{z-1}{\sqrt{(v_r^2 z^2 + v_f^2)}} \right] \left[\frac{(v_f^2 z + v_r^2 z^2)}{(v_r^2 z^2 + v_f^2)^{3/2}} \right] \left(\chi_{1d} + \chi_d z \frac{\bar{F}}{\bar{R}} \right) \tag{10.26}$$

where

$$\Phi'(x) = \frac{1}{\sqrt{2\pi}} \exp\left(-\frac{x^2}{2} \right) \text{ and}$$

$$z = A\bar{R} / \bar{F}$$

z being the safety factor; and v_r and v_f are the variation coefficients of resistance and force, respectively. Equation (10.26) was solved for $v_f = 0.3$, $v_r = 0.1$, $\chi_{id}/\chi_d = 200$, $\chi_d = \alpha$, $\bar{R} = 200$ MPa

$\overline{F} = 200\ kH$. The design parameters are: $z = 1.88$, $\beta = 2.48$, $P_f = 0.0066$, $Ps = 0.9934$.

The same procedure can be used for probabilistic optimization of members under bending. In this case instead of (10.22) we have:

$$\tilde{S} = W\tilde{R} - \tilde{M} \tag{10.27}$$

where W is the section modulus, and M is the bending moment at the cross-section.

10.3 OPTIMIZATION OF PROTECTED STRUCTURES

Protective devices against overloading are now incorporated in many structural systems (Perelmuter, 1999), which influences the reliability parameters and economic efficiency of the considered objects. Failures in such protected structures can be classified as accident-stops or accidents-faults (Zarenin & Zbirko, 1971). Failures become accidents-stops when normal functioning ceases and is not accompanied by structural damage and large financial losses. Accidents-faults are associated with heavy structural damage with considerable financial losses and sometimes casualties. These protective devices are intended to reduce possible accident-faults into accident-stops.

Protection has a dual effect on structural reliability. On the one hand, if the probability of accidental loads is constant, protection would reduce the probability of accidents, as they can occur only when it is off. If the protective device is "absolutely" reliable, accident-faults would be ruled out. On the other hand, the probability of accidents-stops will increase, and more failures will become possible if the protection malfunctions. The efficiency of protective devices is often governed by a weak member, whose failure under load prevents its action on the structure. Safety fuses are a well-known example of this type. Other examples of safety stops are the lifting capacity of cranes, torque-limiting clutches in machinery, safety valves in high-pressure vessels, seismic protection devices, as well as the detachable roofs in explosive factories.

If the efficiency of protective device is perfect and the protected member fails when loaded, then the structure itself needs to be designed only for a load equal to the limit bearing capacity of the member. However, both the efficiency of the protective member and the resistance level of the structure can vary. When the difference between these levels is small, precautions should be taken to ensure that the resistance of the protective device exceeds that of the structure; an increase in this disparity will result in higher cost. Hence the optimal protective level should be determined.

Let's consider a system consisting of two members—a structural member and a protective device—connected in series with regard to resisting an applied load. Based on the formula of overall probability, the non-failure probability, Ps, can be written in the form:

$$P_s = P_{os}\,(1-p_s) + P_{is}p_s \tag{10.28}$$

where ps, P_{os}, and P_{is} are the probabilities of no failure for the protective device, for the unprotected structure, and for the structure with "perfect" protection, respectively. If protection installation leads to $P_{is}=1$, then $P_s = P_{os} + p_s - p_s P_{os}$. This expression corresponds to a parallel connection of the members. Thus, the protection plays the role of permanent reserve. Let's denote the resistance levels and protective device of the protected structure R and r, respectively, and assume that they are normally distributed random values with mean values $\bar{R}$ and $\bar{r}$, and standard deviations s_r and s_{rp}. The protective device doesn't work if $r > R$. Its failure probability P_f will be (Huber, 1964):

$$P_f = 0.5 - \Phi\left(\frac{\bar{R} - \bar{r}}{\sqrt{s_r^2 + s_{rp}^2}}\right) \tag{10.29}$$

where $\Phi(z)$ is a Laplace function:

$$\Phi(z) = \frac{1}{2\pi}\int_0^z \exp\left(-\frac{a^2}{2}\right)da$$

Denoting the variation coefficients for the structure and the protective device by $v_r = s_r/\bar{R}$ and $v_{rp} = s_{rp}/\bar{r}$, taking for simplicity $v_r = v_{rp}$ and denoting also $x = \bar{r}/\bar{R}$, Equation (10.29) can be rewritten as:

$$P_f = 0.5 - \Phi\,[(1 - x)/v_R\sqrt{(1 + x^2)}] \tag{10.30}$$

When the load is a normally distributed random process $F(t)$ with mean value $\bar{F}$, correlation function $\rho_f(\tau)$ and deviation $s_f^2 = \rho_f^2(0)$, then the probability of load $F(t)$ exceeding resistance R in an infinitesimal time interval (density of overshooting) is given by the well-known formula (Rzhanitsyn, 1978):

$$I_r = \frac{1}{\sqrt{2\pi}}\left[\frac{-s_f^2(0)''}{s_R^2 + s_F^2}\right]\exp\left[-\frac{\left(\bar{R} - \bar{F}\right)^2}{2\left(s_R^2 + s_F^2\right)}\right] \tag{10.31}$$

where $s_f^2(0)''$ is a second derivative of the correlation function at $t = 0$.

Using the notations

$$v_f = \frac{s_f}{\bar{F}}, v_f = \sqrt{\frac{-s_f^2(0)''}{\bar{F}}}$$

and introducing safety factor

$$z = \frac{\overline{R}}{\overline{F}}$$

the density of load exceeding the resistance level of protected structure *Ir* can be expressed as follows:

$$I_r = \frac{1}{2\pi}\sqrt{\frac{y_f^2}{z^2 v_r^2 + v_f^2}} \exp \frac{(z-1)^2}{2(z^2 v_r^2 + v_f^2)} \tag{10.32}$$

And for load exceeding the resistance level of protective resistance I_r:

$$I_{rp} = \frac{1}{2\pi}\sqrt{\frac{y_f^2}{x^2 z^2 v_r^2 + v_f^2}} \exp \frac{(z-1)^2}{2(z^2 v_r^2 + v_f^2)} \tag{10.33}$$

Intensiveness of failure for the protected structure is calculated as the probability of simultaneous occurrence of two independent events: $r \geq R$and$F \geq R$ in a time interval *dt*. This situation leads to the expression $I_r P_f$. Intensiveness of functioning for the protective device is $I_{rp}(1 - P_f)$ and for the whole system:

$$I = I_y P_f + I_{rp}(1 + P_f) \tag{10.34}$$

Equation (10.34) represents the probability of the whole system not operating during the time interval *(t, t+dt)*. This information, however, is insufficient for making decision on the behavior of the system, and can even be misleading, as the implications of failure represented by the first and second items in (10.34) are not the same. Thus, it is expedient to estimate cost for comparing accident-faults vs. accident-stops. The overall cost of the whole system *C* is given by:

$$C = C_r + C_{rp} + \int_0^T \left[U_r I_r P_f + U_{rp} I_{rp}(1 - P_f) \right] \xi(t)dt \tag{10.35}$$

where C_r is the construction cost of structure; C_{rp} is the cost of the protective device; U_r is losses due to failure of structure (accidents-faults); U_{rp} is losses due to operation of protective device (accidents-stops); $\xi(t)$=exp(-*nt*) is the discount function, and *n* is the parameter of discount (investment rate). An approximation was proposed, whereby the construction cost is a function of safety factors, such as:

$$C_r = C_{or} z^{\delta},\ C_r = C_{rpo} x^{\varepsilon} z^{\varepsilon} \tag{10.36}$$

TABLE 10.1
Values of δ and ε at different stress states and different proportions of a rectangular cross section of a constant area beam after

Type of loading (stress state)	Values of δ and ε: Changing depth at a constant width	Changing both depth and width	Changing width at a constant depth
Tension	1.000	1.000	1.000
Buckling	-	0.500	1.000
Torsion	1.000	0.667	0.500
Bending	0.500	0.667	1.000

(Driving,1972; Raizer, 198]

where δ and ε depend on stress state under the applied load and cross-section (depth and width) parameters. Table 10.1 presents the value of δ, ε for three different types of cross-section variations and four types of loading of a beam.

It is also assumed that the losses due to a single failure of structure, U_R, can be represented by the cost of its repair plus some additional losses L, so that:

$$U_r = vC_{ro}s^o + L \tag{10.37}$$

Losses from a single failure of protective device include the cost of its replacement plus a fraction of additional losses L:

$$U_{rp} = vC_{rpo}x^{\varepsilon}z^{\varepsilon} + \mu L \tag{10.38}$$

where μ depends on ratio of protective device replacement time to that of structure repair.

It is often possible to take $\mu = 0$, especially where failure of structure is associated with heavy ecological damage or with other losses unrelated to the system's downtime.

Denoting $\eta = L/C_{ro}$, $\theta = C_{rpo}/C_{ro}$, and after integration of (10.35), the overall cost of a system of system can be expressed as

$$C = C/C_{ro} = z^{\delta} + \theta x^{\varepsilon}z^{\varepsilon} + [P_f(vz^{\delta} + \mu)\,I_r + (1 - P_f)\,(\theta x^{\varepsilon}z^{\varepsilon} + \eta\mu)I_{rp}]\;\{[1 - \xi(t)]/n\} \tag{10.39}$$

The relationship of C and parameters x and z is shown in Figure 10.2. Minimization of Equation (10.39) yields the optimal value of safety factor z*, relative level of resistance for protective device x^* and optimal cost C^*. Calculations were made for a structural failure under a variety of parameters T, η, θ, and μ.

For example (Perelmuter, 1999), at $v_r = 0.1$, $v_f = 0.2$, and taking the correlation function of loads $k_f(\tau) = y_f exp(-\chi^2\tau^2)$, with parameter $\chi = 1 year^{-2}$ one can get $y_f = 1.414v_f$. The stress state effect was taken according to Table 10.1 when $\delta = \varepsilon = 1$.

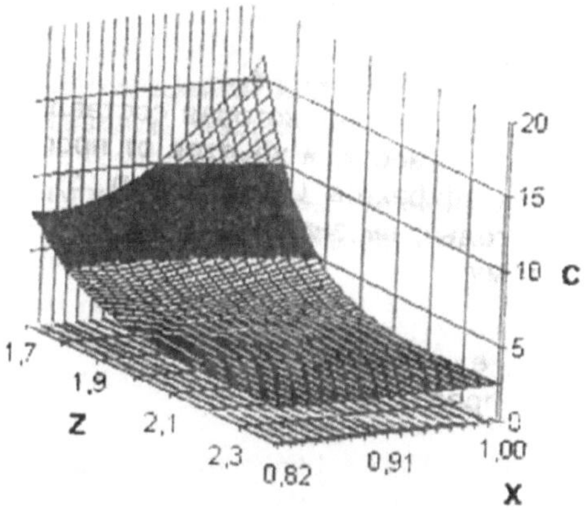

FIGURE 10.2 Functional of cost

Total construction cost *(v = 1)* was included in losses due to structural failure. Results are presented in Table 10.2.

Table 10.2 shows that the optimal value of relative protection resistance, x^*, depends mainly on the factor of outside losses, μ, associated with the accidents-stops. This x^* value remains practically unchanged with variations of T, μ, θ. To determine the optimal safety factor z^0 for unprotected structure one should use the functional:

$$C^0 = z^\delta + (\nu z^\delta + \mu) I_r \frac{1 - \xi(t)}{n} \tag{10.40}$$

Corresponding results are also given in Table 10.2. It follows that recourse to the considered protective devices is justified if they are relatively inexpensive and can be replaced quickly after their failure ($\mu \leq 0.02$).

10.4 OPTIMIZATION OF MULTIPLEX SYSTEMS

General principles and methods of optimization for combined systems with minimum probability of failure are described in (Gnedenko et al., 1969; Geniev, 2000; Raizer, 2009). Results can be applied for solving the optimization problems for different kinds of multiplex systems such as statically determinate frames, continuous beams, and so on. These systems can be considered as chains of elements connected in-series, in which failure of one element entails failure of the whole system. This approach can also be extended to certain statically indeterminate structures where a failure of one element does not mean overall failure but can make further operations difficult. Such a situation should be considered as a warning signal. Consider a system that consists of m separate elements. The following notation is introduced: $V_{oj} = P_{osj}$ is the probability of no failure (or survival) for the j-th element ($j = 1 \div m$), $P_{ofj} = 1 - P_{osj}$ is the probability of failure specified in the

TABLE 10.2
Calculations results for finding optimal value of protection

Parameters				Optimization results (v_r = 0.1, v_f = 0.2, y_f = 0.28, v = 1)					
T	η	θ	μ	x*	z*	C*	C⁰	z*/z⁰	C*/C⁰
10	100	0.001	0.01	0.85	1.87	2.011	2.077	0.949	0.968
			0.02	0.88	1.91	2.062	2.077	0.969	0.993
			0.03	0.90	1.93	2.096	2.077	0.980	≥1
			0.04	0.92	1.94	2.125	2.077	0.985	≥ 1
			0.06	0.93	1.95	2.176	2.077	0.990	≥ 1
			0.08	0.94	1.96	2.223	2.077	0.995	≥ 1
			0.10	0.96	1.96	2.268	2.077	0.995	≥ 1
		0.100	0.01	0.86	1.90	2.020	2.077	0.964	0.972
			0.02	0.89	1.92	2.066	2.077	0.975	0.995
			0.03	0.91	1.93	2.098	2.077	0.980	≥ 1
			0.04	0.92	1.94	2.127	2.077	0.985	≥ 1
			0.06	0.93	1.95	2.177	2.077	0.990	≥ 1
			0.08	0.94	1.96	2.224	2.077	0.995	≥ 1
			0.10	0.96	1.97	2.269	2.077	1.000	≥ 1
	1000	0.001	0.01	0.88	2.15	2.273	2.313	0.973	0.983
			0.10	0.96	2.21	2.532	2.313	1.000	≥ 1
		0.100	0.01	0.88	2.15	2.274	2.313	0.973	0.083
			0.10	0.96	2.21	2.532	2.313	1.000	≥ 1
100	100	0.001	0.01	0.86	1.91	2.042	2.103	0.955	0.971
			0.10	0.95	1.99	2.298	2.103	0.995	≥ 1
		0.100	0.01	0.86	1.91	2.049	2.103	0.995	0.974
			0.10	0.95	1.99	2.298	2.103	0.995	≥ 1
	1000	0.001	0.01	0.88	2.17	2.301	2.339	0.969	0.983
			0.10	0.96	2.23	2.561	2.339	0.995	≥ 1
		0.100	0.01	0.88	2.17	2.302	2.339	0.969	0.984
			0.10	0.96	2.23	2.561	2.339	0.995	≥ 1

preliminary design; C_{oj} is cost of the *j*-th element. It should be noted that C_{oj} must be determined subject to all relevant rules, taking into account static design, specification of loads, material properties and construction cost. For the combined system, the probability of no failure based on the preliminary design will be:

$$V_o = V_{o1} \bullet V_{o2} \bullet \ldots . V_{oj} \bullet \ldots \bullet V_{om} = \prod_{j=1}^{m} V_{oj} \tag{10.41}$$

and the probability of failure for the whole system will be equal to: $P_{of} = 1 - V_o = 1 - P_{os}$.

For real values of V_{oj} ranging usually within $V_{oj} = 0.95 \div 0.99$, an approximate expression widely used in practical applications is:

$$P_{of} = P_{o1} + P_{o2} + \ldots + P_{om} \sum_{j=1}^{m} P_{oj}\ P_{oj} \qquad (10.42)$$

This expression represents the probability of failure with some reserve of reliability as:

$$\sum_{j=1}^{m} P_{oj} \geq \prod_{j=1}^{m} (1 - P_{oj}) \qquad (10.43)$$

Total cost of construction material will be:

$$C_o = C_{o1} + C_{o2} + \ldots + C_{om} = \sum_{j=1}^{m} C_{oj} \qquad (10.44)$$

Solution of the optimization problem is aimed at possible increase of V_o. It can be done by increasing C_o to add allocations δC, $C = C_o + \delta C$ and distributing them optimally among the elements. Such redistribution guarantees, under Equation (10.42), reaching the maximum value for

$$V = V_1 \bullet V_2 \bullet \ldots . Vj \bullet \ldots \ \ldots \bullet Vm = \prod_{j=1}^{m} Vj \qquad (10.45)$$

where V_j corresponds to C_j. We obtain:

$$C_1 + C_2 + \ldots + C_j + \ldots + C_m = \sum_{j=1}^{m} C_j = C_0 \qquad (10.46)$$

The situation when $\delta C = 0$ will be considered later. For analytical solution of the optimization problem, it was assumed that the relation between probability of failure for j-th element P_j and its cost C_j

$$P_j = P_{fj} = (P_{ofj})^{Cj/Coj} \qquad (10.47)$$

This formula satisfies the condition of coincidence P_{fj} with P_{0fj} when $C_j = C_{oj}$, while

$$V_j = 1\text{-}(1\text{-}V_{0j})^{Cj/Coj} \qquad (10.48)$$

Equation (10.48) approximates the relation between V_j and C_j in the vicinity of point (V_{oj}, C_{oj}) only. The connection between C_{oj} and V_{oj} is based on a preliminary solution. At the same time (10.41) and (10.47) satisfy the limit situations: if $C_j \to \infty$, then $V_j \to 1$; if $C_j \to 0$, then $V_j \to 0$. The effective range of approximation (10.48) is $(0.90 \div 0.99) \leq V_{oj} \leq 1$. The problem solution consists in determining a conventional minimum for function (10.45), taking into consideration (10.48) and (10.46). The well-known method of indefinite Lagrange multipliers can be recommended. The Lagrange function F is expressed as follows:

$$F = P_{o1}^{C_1/C_{o1}} + P_{o2}^{C_2/C_{o2}} + \ldots + P_{om}^{C_m/C_{om}} + \lambda[(C_1 + C_2 + \ldots + C_m) - C_0] \tag{10.49}$$

where λ constitutes the Lagrange multiplier. From the solutions of the Lagrange equations:

$$\frac{\partial F}{\partial C_1} = \frac{\partial F}{\partial C_2} = \ldots = \frac{\partial F}{\partial C_m} \tag{10.50}$$

the final expression for C_j can be presented in the form:

$$C_j = \frac{C_{oj}}{\ln(1/P_{oj})} \left\{ \frac{C_0 + \sum_{j=1}^{m} \frac{C_{oj}}{\ln(1/P_{oj})} \cdot \frac{\ln C_{oj}}{\ln(1/P_{oj})}}{\frac{C_{oj}}{\ln(1/P_{oj})}} - \ln\left[\frac{C_{oj}}{\ln(1/P_{oj})} \right] \right\} \tag{10.51}$$

The failure probabilities for the optimized combined system elements are obtained based on Equations (10.47) and (10.51) and using the no-failure probability for the whole system (10.45). An example of practical importance can be considered for the case of equal probabilities of failure for all elements: $P_{o1} = P_{o2} = \ldots = P_{om} =$ const. In this case the expression (10.51) for cost C_j of the j-th element of the optimized system can be presented in a simple form:

$$C_j = C_{oj}\left(1 + \frac{1}{\ln a}\right) \cdot \left(\frac{\sum C_{oj} \ln C_{oj}}{C_o} - \ln C_{oj} \right) \tag{10.52}$$

Numerical example 1

For a combined system of m=4 elements, its design gives $P_{o1} = P_{o2} = P_{o3} = P_{o4} =$ 0.04 *and* the elements cost: $C_{o1} = 8$, $C_{o2} = 12$, $C_{o3} = 16$, $C_{o4} = 20$ (in conditional monetary units). Under strict limitations on the cost (δC=0, C=C_o), the system is optimized by redistributing *Coj*, which yields the minimal probability of failure for the system – P_{0f}. From Equation (10.52) it follows: $C_1 = 9.521>8$, $C_2 = 12.770>12$, $C_3 = 15.597<16$, $C_4 = 18.110<20$. Based on Equation (10.47) one can get: $P_1 =$ 0.021<0.04, $P_2 = 0.032<0.04$, $P_3 = 0.043>0.04$, $P_4 = 0.054>0.04$. These numerical results conflict, in some degree, with the opinion that more expensive elements should be designed with a higher reliability level than the inexpensive ones. Thus, contradiction is explained by the connections between the elements in the analyzed system when a failure of an arbitrary element (regardless of its cost and reliability level) will provoke failure of the whole system. The overall effect of the redistribution of *Coj* among the elements serve as a characteristic of the whole system as $P_f < P_{of}$.

Numerical example 2

For the same system of four elements, let us consider a design different than that in Example 1, namely: $P_{o1} = 0.05$, $P_{o2} = 0.04$, $P_{o3} = 0.03$, $P_{o4} = 0.02$; and equal initial cost of the elements: $C_{o1} = C_{o2} = C_{o3} = C_{o4} = C_o/4 = 14$. Under the same condition of strict limitation on the system cost ($\delta C = 0$, $C = C_o$) as in Example 1, optimization is done via C_o with a view to the minimum probability of failure for the whole system, P_f. Formula (10.51) yields the following results: $C_1 = 14.237 > 14$, $C_2 = 14.493 > 14$, $C_3 = 13.646 < 14$, $C_4 = 12.623 < 14$. From relations (10.47): $P_1 = 0.038 < 0.05$, $P_2 = 0.035 < 0.04$, $P_3 = 0.032 > 0.03$, $P_4 = 0.029 > 0.02$. In this example a higher cost on structural materials was required for the first and second elements, for which the values $P_{o1} = 0.05$ and $P_{o2} = 0.04$ in the preliminary design proved insufficient.

For combined structures with given geometric dimensions and design schemes, this method guarantees the maximum probability of indestructibility within the constraint of limited total cost of the components.

10.5 OPTIMAL ALLOCATION OF PROTECTIVE RESOURCES OF STRUCTURES

The beginning of the 21st century is characterized by a significant increase in the number of disasters caused by emergency situations. You can also add the possibility of terrorist attacks here. The term "disaster" here is understood as any change in the environment that threatens the safety of people and significantly worsens living conditions (Raizer, 2018).

Protection of structures from such impacts cannot be successfully solved without conducting an analysis of the "efficiency-cost" type, since any resources in reality are limited, which inevitably leads to the problem of their rational distribution. The main problem that arises in mathematical modeling is the giant dimension of the problem.

However, we immediately note that the problem allows a decomposition and hierarchical construction of the model. The entire system of protected objects in a country can be represented by a hierarchical branching structure with an additive global objective function (Ushakov, 2013). The proposed method is based on (Gnedenko & Ushakov, 1995). All aspects of the protection of structures that ensure the safety of people, equipment, and structures can be combined into a single aggregated model that can help those who make decisions at various levels.

Let us assume that there are three distinct layers of objects safety protection: federal, state, and local. It is also assumed that all the necessary input data for the model can be provided by the appropriate experts. Let us introduce the following notations:

$F_i(\varphi_i)$: Subjective probability that an object within the country will be protected against type i emergency exposure under the condition that on a federal level one spends φ_i resources.

$S_i^{(k)}(\sigma_i^{(k)})$: Subjective probability that an object within state k will be protected against type i emergency exposure under the condition that on the level of this particular state one spends $\sigma_i^{(k)}$ resources:

$L_i^{(k,j)}(\lambda_i^{(k,j)})$: Subjective probability that particular object j within state k (denoted as pair "k, j") will be protected against disaster of type i under the condition that one spends $\lambda_i^{(k,j)}$ resources.

$W^{(k,j)}$: "Weight" (or "measure of priority") of object (j, k). This factor is determined by the expert community as a factor of the significance of the structure in the infrastructure of the country, or region, or at the local level. Of course, there is a certain analogy with the introduction to building codes as one of reliability coefficients "importance factor" (Raiser, 1995). Here, however, the problem of creating a unified system for protecting the most critical structures from man-made or natural disasters is discussed, both at the federal, state and local levels.

$G_{k,j}$ – the set of all possible emergency impacts for a particular object (k, j).

n_k – the total number of protected objects in region k.

N – the total number of regions in the country.

At the initial stage, we will consider a single object j located in the region k. Suppose that this object (k, j) can only be affected by certain types of emergency impacts that belong to some set $G_{k,j}$. In this case, the degree of protection at the state level is equal to:

$$F^{(k,j)} = \min \{F_i, i \in G_{k,j}\} \tag{10.53}$$

Now consider the regional level (for region k). Using the same arguments, one can write for the object (k,j) the degree of protection at the regional level:

$$S^{(k,j)} = \min \{S_i, i \in G_{k,j}\} \tag{10.54}$$

Assume that on a local level, object (k, j)'s protection is equal to $L^{(k,j)}$. (Postpone, for a while, how this value was obtained.) Then we can assume that measures of all three layers (federal, state, and local) influence an object independently. The probability of safety of an object (k, j) can be written as:

$$P^{(k,j)} = 1 - (1 - F^{(k,j)}) \cdot (1 - S^{(k,j)}) \cdot (1 - L^{(k,j)}) \tag{10.55}$$

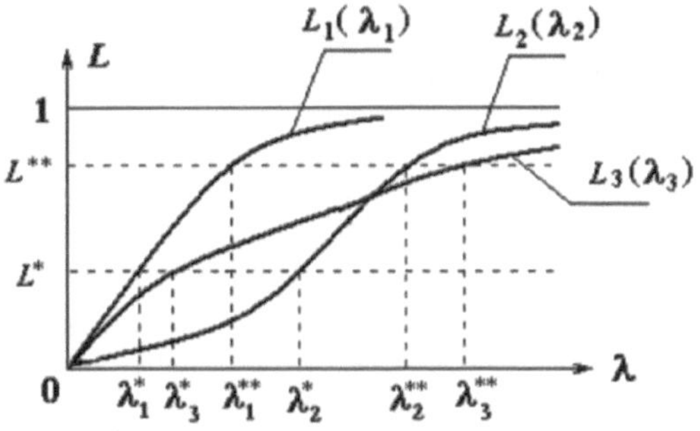

FIGURE 10.3 Examples of possible functions $L_i(\lambda_1)$

Hence, the expected loss, $w^{(k,j)}$, in this case is equal to

$$w^{(k,j)} = W^{(k,j)}\,(1 - P^{(k,j)}) \tag{10.56}$$

Now we move on to calculation of $L^{(k,j)}$ and to the problem of optimal allocation of resources for object (k, j) protection.

Consider $G_{k,j}$, a set of possible accidental actions against object (k, j). On the local layer we know functions $L_i(\lambda_i)$—subjective probability of protection of object (j, k) depending on spent resources λ_i for all possible types of actions, where superscripts (k, j) are omitted, for the sake of simplicity. These functions are presented at Figure 10.3, where for illustration purposes we depict only three such functions. (They should be defined by experts.)

First, consider the direct problem: obtaining the desired level of safety due to measures on the local layer. If the chosen level is L^*, then each of functions $L_1(\lambda_1)$, $L_2(\lambda_2)$, and $L_3(\lambda_3)$ has to have its value not less than L^* because of inequality min $\{L_1(\lambda_1), L_2(\lambda_2), L_3(\lambda_3)\} \geq L^*$ has to be held.

It is obvious that for the minimax criterion to have any $L_i(\lambda_i)$ larger than L^* makes no sense. So the problem of protection resource allocation is solved: the local safety level L^* can be reached if all $L_i(\lambda_i) = L^*$, and in this case one spends a total

$$\lambda^* = \lambda_1^* + \lambda_2^* + \lambda_3^*$$

resources. The number of resources is the minimum for reaching safety level L^*. In an analogous way, if one needs to reach the safety level L^{**}, the expenses related to this level of safety are

$$\lambda^{**} = \lambda_1^{**} + \lambda_2^{**} + \lambda_3^{**}$$

and also are the minimum for this case.

The universe problem (maximization of safety under limited total resources) can be solved with the use of an iterative process of numerical extrapolation. For instance, let total resource λ° be given. One can find two arbitrary solutions of the

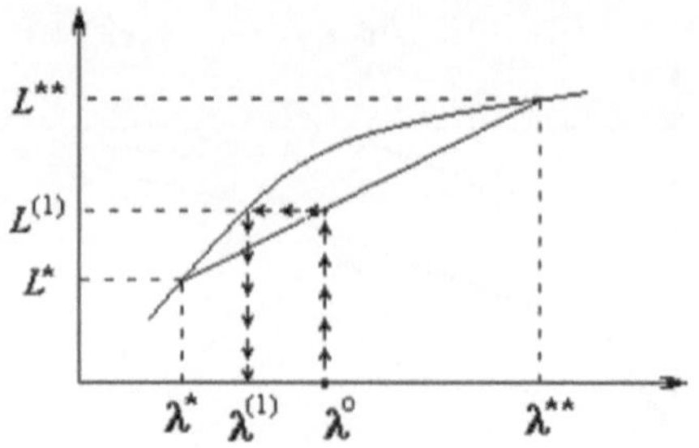

FIGURE 10.4 Illustration of iterative procedure

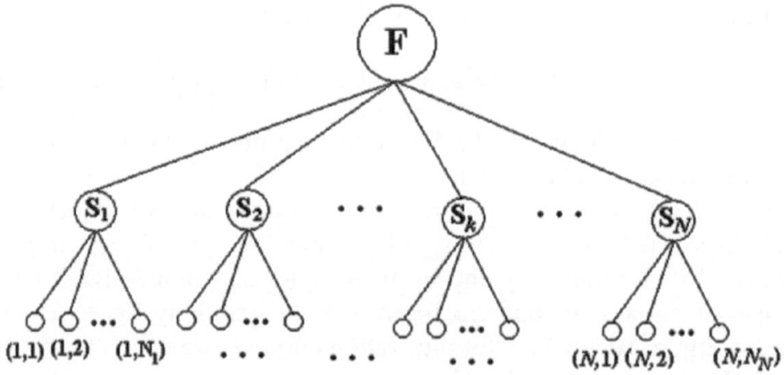

FIGURE 10.5 Three-level branching system

universe problem, say L^* and L^{**} with corresponding values λ^* and λ^{**}. Let all three values satisfy condition

$$\lambda^* \le \lambda^\circ \lambda \le \lambda^{**}$$

Applying linear extrapolation, one finds value $L^{(1)}$ and then, having solved the Inverse Problem for this value, finds a new value λ°, which is used on the second step of the iterative process instead of value λ^*, used at the beginning (see Figure 10.4).

If initially found values λ^* and λ^{**} satisfy conditions $\lambda^* \le \lambda^{**} \le \lambda^\circ$ or $\lambda^\circ \le \lambda^* \le \lambda^{**}$, obviously the iterative process is absolutely similar.

Turning to the aggregated model (Figure 10.5), we note that if an individual efficiency indicator (or the inverse of the damage value) is selected for each object of the lower level of the branching system, then the total efficiency of the system can be found as the sum of these individual indicators.

It follows from probability theory: mathematical expectation of the sum of random variables is equal to the sum of the mathematical expectations of random variables irrespective of dependence of variables.

Indeed, introduce the so-called indicator function of type:
$\delta_{(k,j)} = 1$, if an emergency impact on the object (k, j) is made,

$$\delta_{(k,j)} = 0, \text{ otherwise} \tag{10.57}$$

The random loss for object (k, j) is equal to $\delta_{(k,j)} W^{(k,j)}$ and total random loss of all objects is

$$\sum_{k=1}^{N}\sum_{j=1}^{n_k} \delta_{(k,j)} W^{(k,j)} \tag{10.58}$$

Mathematical expectation of the sum of random variables is defined as

$$w_{\text{Total}}\{F_i, \forall i; S_i^{(k)}, 1 \le k \le N; L_i^{(k,j)}, 1 \le j \le n_k\} =$$

$$E\left\{\sum_{k=1}^{N}\sum_{j=1}^{Nk} \delta_{(k,j)} W^{(k,j)}\right\} = \sum_{k=1}^{N}\sum_{j=1}^{Nk} E\left\{\delta_{(k,j)}\right\} W^{(k,j)}$$

$$= \sum_{k=1}^{N}\sum_{j=1}^{Nk} \left(1 - P^{(k,j)}\right) W^{(k,j)} = \sum_{k=1}^{N}\sum_{j=1}^{Nk} w^{(k,j)} \tag{10.59}$$

where $P^{(k,j)} = 1 - (1 - F^{(k,j)}) \times (1 - S^{(k,j)}) \times (1 - L^{(k,j)})$, and, in turn, these values are defined as

$$F^{(k,j)} = \min\{F_i, i \in G_{kj}\};\ S^{(k,j)} = \min\{S_i, i \in G_{kj}\};\ L^{(k,j)} = \min\{L_i\}.$$

In other words, Equation (10.59) gives the total expected loss, taking into respect the "weight" of each loss.

At the same time, it is easy to calculate the total expenses, C_{Total} on all protection measures on all three layers:

$$C_{Total}\{\varphi_i, \forall i; \sigma_i^{(k)}, 1 \le k \le N; l_i^{(k,j)}, 1 \le j \le n_k\} = \sum_{\forall i} \varphi_i + \sum_{\forall i}\sum_{k=1}^{N} \sigma_i^{(k)} + \sum_{\forall i}\sum_{k=1}^{N}\sum_{i=1}^{n_k} \lambda_i^{(k,j)} \tag{10.60}$$

Having objective Equations (10.59) and (10.60), one can formulate the following optimization problems:

Direct problem. Optimally allocate total available resources that guarantee the minimum possible loss of accidental actions.

Inverse problem. Optimally allocate resources that guarantee the acceptable expected loss of defended objects against accidental actions with minimum possible expenses.

Solution of these problems with the use of steepest descent method is demonstrated on a simple illustrative numerical example with fictitious data concerning expert assessments of the significance factor (priority) and the degree of protection of structures without additional special measures. (These objects can be, for example, a high-rise building, a hydroelectric power station, an offshore oil platform, a stadium, a concert hall, etc.)

Let's assume that we are considering three structures with the following characteristics.

Structure 1: "Weight" of Importance = 10; Level of protection with no special measures $P_1^{(0)} = 0.5$.

Safety	0.9	0.95	0.97	0.99
Expenses	2	4	7	12

Structure 2: "Weight" of Importance = 3; Level of protection with no special measures $P_2^{(0)} = 0.8$.

Safety	0.9	0.95	0.97	0.99
Expenses	1	2	4	8

Structure 3: "Weight" of Importance = 7; Level of protection with no special measures P3(0) = 0.9.

Safety	0.9	0.95	0.97	0.99
Expenses	0.5	1	2	5

The "significance factor" may depend on the number of people, the cost of the consequences, and also on the national importance of the structure. One has to calculate "discrete gradients" (relative increments) for each structure k according to the formula:

$$\gamma_k^{(s)} = W_k \frac{P_k^{(s)} - P_k^{(s-1)}}{C_k^{(s)} - C_k^{(s-1)}} \qquad 10.61)$$

where W_k is the "significance factor" of the structure k,

$P_k^{(s)}$is the level of security at step s of the process of increasing the security of this structure, and

$C_k^{(s)}$are expenses related to the level of protection at step s of the process of defense improving.

Let us construct Table 10.3, which will be used (in very simple way) to get an optimal allocation of money for defenses of all three structures.

The numbers given in Table 10.4 show in which order it is most rational to increase the safety of structures. The final results are shown below:

TABLE 10.3
Values of step-by step "gradients"

№ step	Value of "gradient" step-by-step: Structure1	Structure 2	Structure 3
1	$10 \cdot \frac{0.9-0.5}{2} = 2$	$3 \cdot \frac{0.9-0.8}{1} = 0.3$	$7 \cdot \frac{0.95-0.9}{1} = 0.35$
2	$10 \cdot \frac{0.95-0.9}{4-2} = 0.25$	$3 \cdot \frac{0.95-0.9}{2-1} = 0.15$	$7 \cdot \frac{0.97-0.95}{2-1} = 0.14$
3	$10 \cdot \frac{0.97-0.95}{7-4} = 0.067$	$3 \cdot \frac{0.97-0.95}{4-2} = 0.03$	$7 \cdot \frac{0.99-0.97}{5-3} = 0.07$
4	$10 \cdot \frac{0.99-0.97}{12-7} = 0.004$	$3 \cdot \frac{0.99-0.97}{8-4} = 0.0015$	*

TABLE 10.4
Ordered values of "gradients"

№ шага	Value of "gradient" step-by-step: Structure 1	Structure 2	Structure 3
1	1	3	2
2	4	5	6
3	8	9	7
4	10	11	*

(1) Initial expected loss (no protection) is equal to

$$w^{(0)} = W_1 \cdot (1 - P_1^{(0)}) + W_2 \cdot (1 - P_1^{(0)}) + W_3 \cdot (1 - P_1^{(0)})$$
$$= \cdot 10 \cdot 0.5 + 3 \cdot 0.2 + 7 \cdot 0.1 = 3.8$$

(2) After the 1st step (the protection of only Structure 1 improves, which has the highest value of the relative increment of the safety function), the total losses are equal to

$$w^{(1)}_{Total} = W_1 \cdot (1 - P_1^{(1)}) + W_2 \cdot (1 - P_1^{(0)}) + W_3 \cdot (1 - P_1^{(0)})$$
$$= \cdot 10 \cdot 0.1 + 3 \cdot 0.2 + 7 \cdot 0.1 = 1.8$$

and the spent resources are equal to $C^{(1)} = 2$

(3) After the second step the total expected loss is equal to

$$w^{(2)}_{Total} = W_1 \cdot (1 - P_1^{(1)}) + W_2 \cdot (1 - P_1^{(1)}) + W_3 \cdot (1 - P_1^{(0)})$$
$$= \cdot 10 \cdot 0.1 + 3 \cdot 0.1 + 7 \cdot 0.1 = 1.5$$

and the spent resources are equal to $C^{(2)} = 2 + 1 = 3$

(4) After the third step the total expected loss is equal to

$$w^{(3)}_{Total} = W_1 \cdot (1 - P_1^{(1)}) + W_2 \cdot (1 - P_1^{(1)}) + W_3 \cdot (1 - P_1^{(1)})$$
$$= \cdot 10 \cdot 0.1 + 3 \cdot 0.1 + 7 \cdot 0.05 = 1.15$$

and the spent resources are equal to $C^{(3)} = 2 + 1 + 1 = 4$

(5) After the fourth step the total expected loss is equal to

$$w^{(4)}_{Total} = W_1 \cdot (1 - P_1^{(2)}) + W_2 \cdot (1 - P_1^{(1)}) + W_3 \cdot (1 - P_1^{(1)})$$
$$= \cdot 10 \cdot 0.05 + 3 \cdot 0.1 + 7 \cdot 0.05 = 0.9$$

and the spent resources are equal to $C^{(4)} = 2 + 1 + 1 + 2 = 6$

(6) After the fifth step the total expected loss is equal to

$$w^{(5)}_{Total} = W_1 \cdot (1 - P_1^{(2)}) + W_2 \cdot (1 - P_1^{(2)}) + W_3 \cdot (1 - P_1^{(1)})$$
$$= \cdot 10 \cdot 0.05 + 3 \cdot 0.05 + 7 \cdot 0.05 = 0.75$$

and the spent resources are equal to $C^{(5)} = 2 + 1 + 1 + 2 + 1 = 7$

(7) After the sixth step the total expected loss is equal to

$$w^{(6)}_{Total} = W_1 \cdot (1 - P_1^{(2)}) + W_2 \cdot (1 - P_1^{(2)}) + W_3 \cdot (1 - P_1^{(2)})$$
$$= \cdot 10 \cdot 0.05 + 3 \cdot 0.05 + 7 \cdot 0.03 = 0.61$$

and the spent resources are equal to $C^{(6)} = 2 + 1 + 1 + 2 + 1 + 1 = 8$

(8) After the seventh step the total expected loss is equal to

$$w^{(7)}_{Total} = W_1 \cdot (1 - P_1^{(2)}) + W_2 \cdot (1 - P_1^{(2)}) + W_3 \cdot (1 - P_1^{(3)})$$
$$= \cdot 10 \cdot 0.05 + 3 \cdot 0.05 + 7 \cdot 0.01 = 0.47$$

and the spent resources are equal to $C^{(7)} = 2 + 1 + 1 + 2 + 1 + 1 + 2 = 10$

(9) After the eighth step the total expected loss is equal to

$$w^{(8)}_{Total} = W_1 \cdot (1 - P_1^{(3)}) + W_2 \cdot (1 - P_1^{(2)}) + W_3 \cdot (1 - P_1^{(3)})$$
$$= \cdot 10 \cdot 0.03 + 3 \cdot 0.05 + 7 \cdot 0.01 = 0.37$$

and the spent resources are equal to $C^{(8)} = 2 + 1 + 1 + 2 + 1 + 1 + 2 + 3 = 13$

(10) After the ninth step the total expected loss is equal to

$$w^{(9)}_{Total} = W_1 \cdot (1 - P_1^{(3)}) + W_2 \cdot (1 - P_1^{(3)}) + W_3 \cdot (1 - P_1^{(3)})$$
$$= \cdot 10 \cdot 0.03 + 3 \cdot 0.03 + 7 \cdot 0.01 = 0.31$$

and the spent resources are equal to $C^{(9)} = 2 + 1 + 1 + 2 + 1 + 1 + 2 + 3 + 2 = 15$

(11) After the tenth step the total expected loss is equal to

$$w^{(10)}_{Total} = W_1 \cdot (1 - P_1^{(3)}) + W_2 \cdot (1 - P_1^{(3)}) + W_3 \cdot (1 - P_1^{(3)})$$
$$= \cdot 10 \cdot 0.01 + 3 \cdot 0.03 + 7 \cdot 0.01 = 0.21$$

and the spent resources are equal to $C^{(10)} = 2 + 1 + 1 + 2 + 1 + 1 + 2 + 3 + 2 + 5 = 20$

The process of constructing tradeoff cost-protection can be continued. Graphical presentation on the steepest descent solution is shown in Figure 10.6.

Dependencies of this kind can be used both for the steepest descent method and for constructing the dominant sequence when using the algorithm (Kettelle, 1962). The next phase of the study will have to include an aggregated model that includes a set of objects selected to ensure their safety in the event of a man-made disaster.

Naturally, the implementation of the aggregated mathematical model is possible only with the use of computers due to the huge dimension of the problem. The presence of a computer model will allow us to consider more realistic statements, take into account a greater number of different factors, introduce vector characteristics of costs (human resources, funds, material resources), as well as conduct a scenario analysis of various situations. This method easily allows you to include new factors, new scenarios, and working in the "what-if" mode will

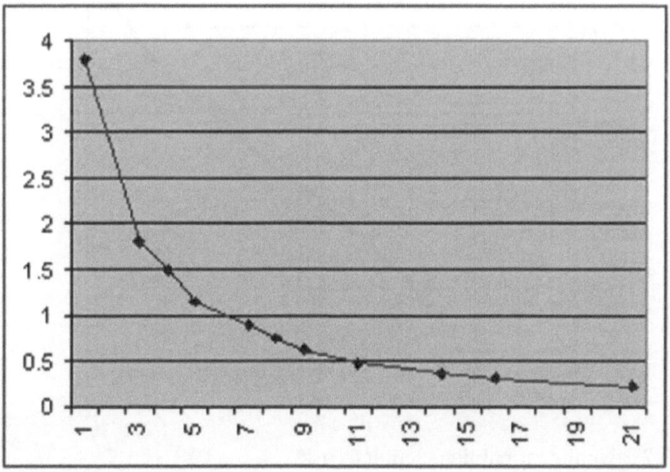

FIGURE 10.6 Tradeoff cost-protection

allow you to find the pros and cons of various strategies for ensuring the safety of structures under man-made, as well as natural and other emergency, impacts.

10.6 MATHEMATICAL MODEL OF PUBLIC OPINION

When failure of a structure or a transportation vehicle causes a serious accident, especially in cases of injuries and fatalities, its designers and certifying/regulatory agencies usually find themselves in the spotlight exposed to the vagaries of public opinion. Moreover, every such accident evokes a negative reaction to the failed structure, a reaction that is aggravated if these failures recur in different places more than two or three times. It should also be noted that the actual severity of the accident is not necessarily a decisive factor in the reaction. For example, three accidents closely successive in a population of one million similar structures can have a stronger impact than a single accident of a similar structure in a population of only 1,000, although their probabilities are glaringly disproportionate – 1:1000 versus 3:1000000.

We see here an obvious deviation from the principle of equal probability for structures with equal responsibility, if they are realized in different social environments. Another obvious anomaly is the observed fact that public reaction is insensitive to the size of the human population at risk: it is practically the same regardless of the casualty count (satiation effect). Sinitsyn (1985) indicates a nonlinear relation between risk values and the meaning of "profit" (Figure 10.11).

It is also important to consider some peculiarities in the evolution of public reaction (Elochin, 1994). Assume that there is a conveyance of the story of an accident from those "who knows" to those "who don't know yet." The smaller the first

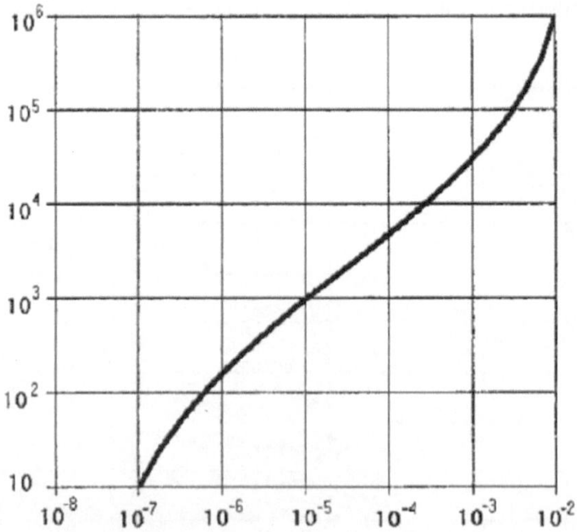

FIGURE 10.7 Nonlinear public reaction to risk

category, the relatively larger is the potential second. However, as the latter shrinks with time and the former grows, the distribution rate is slowed down. Thus, the rate of growth of the informed category C can be presented in as follows:

$$\frac{dC}{dt} = \alpha C(1 - \frac{C}{N}) \tag{10.62}$$

where N is the size of the potential audience, α is a constant, and t is time. The solution of differential equation (10.62) represents the well-known logistic curve for growth of large populations (Arnold, 1997) and reads:

$$C = \frac{1}{1 + \left(\frac{N}{C} - 1\right)\exp[-\alpha(t - t_0)]} \tag{10.63}$$

where C_0 is the number of initial "who knows" at $t = t_0$. For simplicity we will take $t_0 = 0$. The curves are plotted in Figure 10.12 (Perelmutter, 1999) for two values of non-dimensional parameter $k = C_0/N$; $k = 0.1$ and $k = 0.01$ with $\alpha = 2$ in both cases.

The decrease of intensity in public reaction R with time can be formulated as:

$$R(\tau) = R_0 \exp[-\psi(\tau - t)] \tag{10.64}$$

where coefficient ψ denotes the relaxation rate. Thus, for an "elementary" group $dC(t)$ who learned about the accident at moment t, the reaction can be expressed as:

$$dCR(\tau) = A\exp\left[\frac{-\psi t}{1 + \left(N / C_0 - 1\right)\exp(-\alpha t)}\right]^2 \tag{10.65}$$

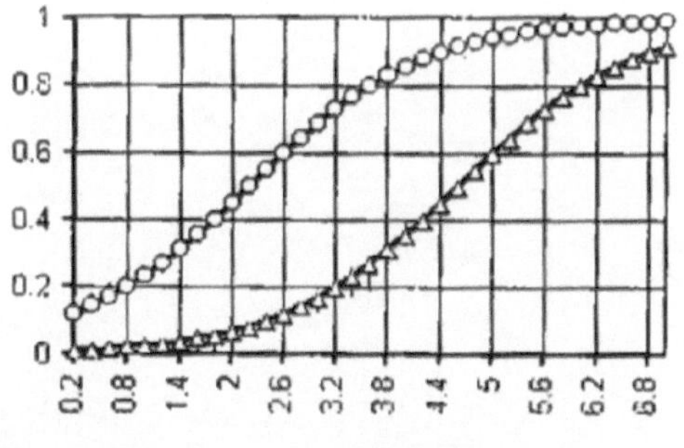

FIGURE 10.8 Growth of the number of "who knows" (Circles-k=0.1, triangles-k=0.01); X-coordinate-t (years), Y-coordinate-degree of satiation

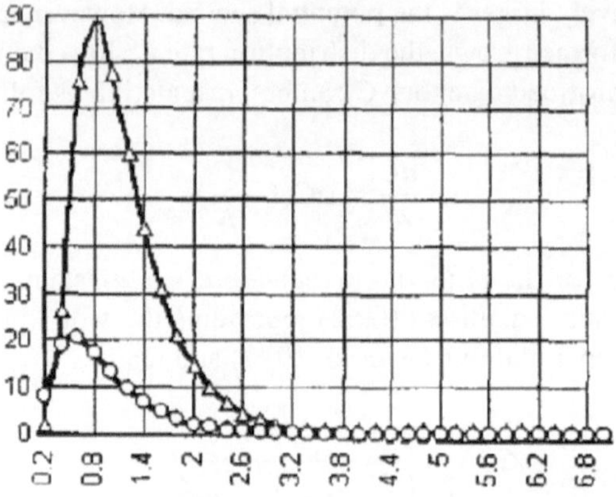

FIGURE 10.9 Public concern vs. time after catastrophic event (circles-$k = 0.01$, triangles-$k = 0.1$); *X*-coordinate-*t* (years) after an accident; *Y*-coordinate-conditional public malaise

where $A = \alpha CR_0 N(N / C_0 - 1)\exp(-\psi\tau)$; and the overall public reaction U at moment τ is as follows:

$$U(\tau) = A\int_0^\tau \frac{\exp(\psi / t)}{[1 + (N / C_0 - 1)\exp(-\alpha t)]^2} dt \qquad (10.66)$$

Curves for conditional values of $U(\tau) / NR_0$ are shown in Figure 10.13 at $\psi/\alpha = 0.5$ for the same *k* as in Figure 10.12.

It is seen that the initial trend of increasing the public concern is subsequently reversed. After a while the cumulative effect of public concern diminishes approaching to zero.

11 Natural Disasters and Structural Survivability

Coincidences, in general, are great stumbling blocks in the way of that class of thinkers who have been educated to know nothing of the theory of probabilities—that theory to which the most glorious objects of human research are indebted for the most glorious of illustrations.

Arsene Dupin, in "The Murders in the Rue Morgue,"
Edgar Allen Poe (1809–1849)

11.1 DISASTER PREDICTION PROBLEMS

The term "disaster" is understood here as any environmental change endangering human lives and materially deteriorating living conditions. A considerable part of disasters comprises natural calamities. Disasters can originate both inside the Earth due to tectonic processes (earthquakes, volcanic eruptions) and near or on its surface due to atmospheric processes (floods, tsunamis, hurricanes, tornados, land and mud sliding, avalanches, karsts, ground heaves and settlements). In many cases successions of interdependent disasters are possible, including those occurring in different media (earthquake-tsunami, earthquake-landslide, and hurricane-flood, etc.). Analysis of conditions associated with the onset and the development of dangerous natural processes is the subject of both natural research and engineering study. The mechanisms of dangerous natural phenomena can be described by direct cause–effect relations. Predicting the type, time, and magnitude of an expected disaster, if feasible, can only be probabilistic. Therefore, using a probabilistic approach and a reliability theory appears to be the most efficient and the only practical tool for analyzing structures operating in areas where natural calamities can be expected.

The level of understanding the origination of natural disasters and, hence, the efficiency in predicting their time, conditions, and severity lags behind the practical needs of national economy. And so does the development of measures for their prevention and mitigation. This is partly due to the absence of common approaches for modeling some natural disasters and methods of their prediction. The forecast of rare disastrous events can be based either on physical models of the phenomenon in question, or using statistical methods, of both. For the latter, statistical

DOI: 10.1201/9781003265993-11

information for a rather long period of time is needed. Practically, however, the prediction is usually based on limited information, often imprecise, and sometimes merely incorrect.

The accuracy of statistically based forecasts fluctuates within a certain range. Different methods should also be used in prediction. The methods commonly used for sufficiently substantiated statistically based predictions (Cramer, 2003; Gross, 2005; Aitkin & Taroni, 2004; Bedford & Cooke, 2001; Calafiore et al., 2006) include: multi-dimensional regression analysis, theory of quantitative analysis, graph theory for error analysis, Delphi method (method of expert evaluation), and statistical analysis. In addition to these five approaches, forecasting the disastrous events and preventing the maximum risk and losses due to abnormal actions in recent decades has been based on the theory of fuzzy sets (Klir & Yuan, 1995). Its application is due to the fact that any classification, any rule of decision making, any model (theoretical or calculated) can be correlated with its fuzzy analogue. For example, classification implies the breakdown of a multitude of elements into classes or groups of similar elements. A prescriptive classification refers each element to a single definite class, whereas in a fuzzy classification each element can belong to different classes depending on certain conditions. The fuzzy classification is generally more realistic than the prescriptive one. The use of the theory of fuzzy sets makes it possible to develop optimal solutions in accordance with practical restrictions.

11.2 STATISTICAL EVALUATION OF NATURAL DISASTERS

Unlike traditional methods, the probabilistic approach makes it possible to evaluate a possible magnitude of disastrous actions in a particular area when analyzing both structures and soils. Therefore, when elaborating a probabilistic concept for natural disasters one should primarily consider in a general form the feasibility of using the statistical approach for representing the disastrous effects. The statistical analysis for the problem should be focused on a probabilistic prediction of the time, location, and magnitude of a natural disaster, or alternatively, on the occurrence probability of a particular type of disaster at the given service life and location of the structure during the given time period. Besides probabilistic predictions, a direct forecasting based on warning signs can also be used. Reliable warning signs, however, are often detected just before the disaster and cannot be taken into account in a long-term prediction influencing the engineering solution.

Statistical methods are used to have a prior notion of the frequency and extent of disasters possible in a particular region. Observations for previous years can give the information on frequency and other parameters of the past disasters. The future characteristics can be estimated by extrapolating the probability of the past events. The estimate, however, can be rather conventional since the acquired statistical data refer to a limited time range. Therefore, the data processing should be based on specially developed statistical models whose physical adequacy to the phenomena under consideration would make the extrapolation trustworthy. Since natural disasters are extreme occurrences (earthquake or/and tsunami of high intensity, etc.), they can be characterized by the "statistics of rare phenomena." Extreme values are usually associated with small occurrence probabilities. The Poisson's distribution appears to be most appropriate in this case for modeling the time character of disasters by the Poisson's process.

This law determines the number of occurrences of rare events, while the theory of extreme values (Gumbel, 1967) considers their values. For random events where extreme values play major role, the asymptotic theory of extreme order statistics provides in some cases relatively accurate but mostly approximate probabilistic models. Therefore, if the basic assumptions of the model have a similarity to the main conditions of a real situation of a catastrophic action, the complex real conditions can be simulated by a considerably simple asymptotic model. Areas, where dangerous phenomena can occur at intensity levels exceeding those on the record (earthquake exceeding the design level, etc.), can be determined and assessed by the test's observations of similar but less intensive occurrences.

A specific feature of natural disasters (as well as human-caused disasters) is that they are virtually unavoidable. Natural disasters are characterized by power and uncontrollability. Typical of human-caused events is that they result from quick development of modern technologies and their products. These factors and the human factor constitute a weak link in a chain of events leading to tragic consequences (Chernobyl or Fukushima accidents for example). The main task here is to predict possible disasters, localize them, and mitigate possible losses. Design of any structure should be preceded by the analysis of all possible types of natural or human-caused disasters in terms of the occurrence probability, secondary disasters, feasibility of localization, preventive measures not connected with the design methods, and the possible damage.

11.3 SAFETY CRITERIA OF UNIQUE STRUCTURES

Prior to discussing the safety criteria, we should clarify the notion of a unique structure and natural or other effects that, determining its vulnerability, are detrimental for human health. The notion of the structural uniqueness and that of the danger of natural or other calamities are interconnected. Considering the structural safety in terms of the danger to human life and health, we should not connect the uniqueness of the structure with its cost or with the expected material losses. The uniqueness should as well be linked with the level of the danger to people, regardless of its probability and of factors causing it, such as the structure's operational profile, size, how it has been built, the presence of radioactive products, and so on. Hence, unique structures are those whose damage or collapse, no matter how low its probability could be, endangers the human life and health either inside or, what is more often, outside the structure.

The foregoing definition of structural uniqueness allows one to refer to such structural projects of national economy (energy, transportation, and others) including social sphere, whose damage and collapse would endanger human life and health. Vulnerability of unique structures exposed to disastrous actions and possibility of their damage or collapse depend on:

- the extent to which the loads due to disastrous events exceed standard loads;
- the influence of secondary factors (explosions, fires) due to disastrous events;
- the errors occurred in the design, analysis, and in choosing the location of the structure, both during its construction and maintenance;
- poor workmanship, inadequate structural materials, and outdated standards for materials, construction and maintenance.

When analyzing structural vulnerability or safety it is expedient to single out critical elements on which the structural safety depends most. For many structures these are the load-bearing members that determine the structural strength and stability (foundation, columns, floors, joints, supports, etc.), as well as the members resisting explosion or fire caused by disastrous events and ensuring reliable operation of safety systems. For a number of unique buildings the critical elements are associated with the protection from radioactivity and with radiation safety.

Different characters of the critical elements require performing, when choosing safety criteria of unique units, a systematic analysis in order to identify these elements and to assess the consequences of their failure. The systematic analysis of structural safety should include elaborating the critical scenarios of the event in question taking into account its possible development, the structure's uniqueness, the presence and type of the critical elements, the consequences of their failure, the implications for human life and for the environment, and so on. Generally, every natural phenomenon and every unique structure requires a scenario allowing one to take their specificity into account and to obtain statistical data for generalizing the consequences. The elaboration and analysis of the scenarios require a great professional skill and effort of people acquainted with the task.

To specify qualitative and quantitative safety criteria of unique structures exposed to any types of natural effects, an integrated approach should be based on:

- systematic deterministic analysis of scenarios describing how the disastrous event affects the unique structure revealing particular quality criteria;
- probabilistic risk analysis determining particular and general probabilistic safety criteria that include those for the limit states representing the extent of failure, and criteria for endangering human life and health;
- cost-benefit analysis for assessing the safety based on optimization of investments for protection against unfavorable effects with due regard for socio-economic factors.

11.4 SURVIVABILITY OF STRUCTURAL SYSTEMS

Different situations that had not been foreseen during the design of a structure can occur as a result of natural or human-caused abnormal actions. These situations can be classed according to failure type, degree of damage, and final state. The following types of failure can be considered for the ultimate limit state:

- loss of strength in time of plastic, brittle, or fatigue failure of elements;
- elastic or inelastic buckling of structural elements;
- loss of the stable equilibrium of the whole structure.

According to the degree of its intensity the damage can include the following:

- Rapidly progressive failure of the whole structure. This form of failure is typical for brittle failure when a damage of separate elements can cause dynamic effects in other elements of a structure (Chapter 6).
- Gradually growing failure of an accidental character as a result of plastic deformations. This situation will stop operation and require repair. This form of failure is typical for structures made of ductile materials when a failure of separate element(s) is accompanied by growing displacements and redistributions of inner forces.

It should be noted that in failure analysis of structures, the failure process has an avalanche-like character of a sequence of failures of elements of a multi-member system. And the notion of damage can be applied to both partial failure and complete collapse of the system. In a majority of cases, however, a number of individual (or partial) failures do not necessarily lead to a total breakdown. This is usually due to a redistribution of stresses which typically occurs in statically indeterminate systems, so the structure continues to operate, though at a lower efficiency and redundancy. The situation can be characterized by the reserves of load-bearing capacity of the structures. The safety margins are usually implicitly embedded in the design, based on experience or engineering judgment. To provide a reasonable reliability level the structure should be designed in such a way that the zone, where damage to main load-bearing elements is likely to occur, should be localized and isolated (Lew, 2005).

The term *survivability* of a structural system is understood as the ability to operate in the period of disastrous events and to prevent progressive failures. Survivability is an important and, especially for a unique structure, indispensable property, since its reliable performance is possible only if an appropriate level of survivability is ensured. It appears natural that a quantitative assessment of this ability should be based on a probabilistic approach, similar to what has been conventional for structural reliability evaluation. Following this logic, the level of survivability can be determined by a probability of certain events characterizing the process of failure. It is logical to consider how a critical state is attained in the process of successive failures of members. This can be the failure of a number of members and formation of an instantaneous mechanism, or the failure of isolated members, and so on. According to this approach, a structure, having been in a damaged condition, can be considered capable to survive if the probability of that event is not so high, as compared to its undamaged condition (other criteria can be used as well).

The index of survivability (Raizer, 2009) can be defined using the following notion: invariability of a system can be defined as a logical function of n logical variables $Y = Y(x_1, x_2, \dots x_n)$, which take the value $Y = 1$ for invariable and $Y = 0$ for variable system (Cherkesov, 1987).

Then the index of survivability can be expressed as:

$$n_y = \left(1-\frac{1}{n}\right)^{n-1} \frac{1}{n}\sum_{i=1}^{n} Y_i + \left(1-\frac{1}{n}\right)^{n-2} \frac{1}{n^2}\sum_{j\le i}^{n}\sum_{i=1}^{n} Y_{ij} + \left(1-\frac{1}{n}\right)^{n-3} \frac{1}{n^3}\sum_{j\le i}\sum_{i\le k}\sum_{k=1}^{n} Y_{ijk} + \dots \tag{11.1}$$

where:

$$Y_i = [Y\{x_1, \dots, x_n\}; \quad x_i = 1 \dots x_s = 0 \;\; (i \ne s)]$$

$$Y_{ij} = [Y\{x_1, \dots, x_n\}; \quad x_i = 1,\; x_j = 1, \dots x_s = 0 \;\; (ij \ne s)]$$

$$Y_{ijk} = [Y\{x_1, \dots, x_n\}; \quad x_i = 1,\; x_j = 1,\; x_k = 1 \dots x_s = 0 \;\; (ijk \ne s)]$$

The number of items in (10.1) does not exceed the degree of static indeterminacy r. As the coefficients before sums

$$K_j = \left(1-\frac{1}{n}\right)^{n-j} \frac{1}{n^j}$$

are related as

$$\frac{K_{j+1}}{K_j} = \frac{1}{n-1}$$

Then the value of each consequent item is smaller than that of the foregoing.

Let us consider the formal way for calculating the terms in the sum (10.1) based on the finite element analysis of the structure (Perelmuter, 1995). The finite elements are characterized with a matrix of generalized internal displacements (deformations) **Z** and a matrix of corresponding internal forces (stresses) **S**. They are related as follows:

$$\mathbf{S} = \mathbf{FZ} \tag{11.2}$$

where F is the definite, symmetric, positive matrix.

Internal displacements **Z** are related with external (joints') displacements **u** with the deformations compatibility condition:

$$\mathbf{Z} = \mathbf{Q}^T\mathbf{u} + \mathbf{d} \tag{11.3}$$

Stresses S are satisfied with the condition

$$\mathbf{QS} = \mathbf{p} \tag{11.4}$$

where **d** and **p** in (11.3) and (11.4) are external dislocations and actions; and $\mathbf{Q}^T$ means transposed (conjugated) matrix **Q**. The dimension of the vectors **S**, **Z**, **d** is equal to the sum of the internal unknowns m. The dimension of the vectors u, p is equal to the common number of external joints unknowns n. For invariable structures the rectangular matrix $\mathbf{Q}(n*m)$ has the rank n, while $r = m$, where n denotes the degree of static indeterminacy. Unknowns **S** and **Z** are eliminated, as is usually done in the method of displacements. The equations for joint's displacements are

$$(\mathbf{QFQ}^T)\mathbf{u} = \mathbf{p} - \mathbf{QFd} \tag{11.5}$$

From (11.5)

$$\mathbf{u} = \frac{\mathbf{p} - \mathbf{QFd}}{\mathbf{QFQ}^T} \tag{11.6}$$

With no external loads ($\mathbf{p} = 0$), forces (stresses) can be expressed as

$$\mathbf{S} = \mathbf{F}(\mathbf{Q}^T\mathbf{u} + \mathbf{d}) = \mathbf{Fd} + \frac{\mathbf{FQ}^T}{\mathbf{QFQ}^T} - \mathbf{QFd} = \left(1 - \frac{\mathbf{FQ}^T}{\mathbf{QFQ}^T}\mathbf{Q}\right)\mathbf{Fd} = (1 - \mathbf{M})\mathbf{Fd} \tag{11.7}$$

Here $\mathbf{M} = \mathbf{FQ}^T(\mathbf{QFQ}^T)^{-1}\mathbf{Q}$ and $\mathbf{N} = 1 - \mathbf{M}$ are projection matrixes. It follows from the conditions that $\mathbf{MM} = \mathbf{M}$ and $\mathbf{NN} = \mathbf{N}$. It is known that connecting with the full rank matrix **Q**, the projector matrix $\mathbf{R} = 1 - \mathbf{Q}^T(\mathbf{QQ}^T)\mathbf{Q}^{-1}$ will transform any vector $\mathbf{d}^0$ into the vector $\mathbf{S}^0 = \mathbf{Rd}^0$ which satisfies the condition $\mathbf{QS}^0 = 0$. As **Q** is the matrix of a system of equilibrium equations, then $\mathbf{S}^0$ can be interpreted as a vector of pre-tension forces being compatible with zero external load. Here $\mathbf{d}^0$ can also be interpreted as a vector of arbitrary dislocations disturbances, (extension of a truss bar, for example) which have induced the pre-stressing for stresses $\mathbf{S}^0$. It can be shown that **R** is compatible with **N** if all stiffness parameters are taken such that $\mathbf{F} = 1$. This is acceptable in the case when only structural characteristics of the design model are analyzed.

If there are absolutely necessary elements in the system, that is, such elements whose removal would result in a loss of geometric invariability, then it is impossible to create the pre-stressing for any dislocations $\mathbf{d}^0$. It follows that the matrix R will contain totally zero lines and columns, which correspond to these elements. As components of the projection matrix **N** have the meaning of forces in the elements due to unity values of elongations $\mathbf{d}^0$, then the matrix **N** or **R** can be easily obtained using FEM program. For this it is necessary to analyze the structure step by step at n versions of dislocations due to each single action. The force vector in the elements of the system corresponding to the jth version of loading comes to the jth column of the projector. If the kth element is excluded from the system then it is necessary that the condition $(\mathbf{S}^0)_k = 0$ can be preserved under any actions as the force in the excluded element will not appear. To reach this case it is necessary to exclude the variable $(\mathbf{S}^0)_k$ from the system of equations $\mathbf{S}^0 = \mathbf{R}\mathbf{d}^0$. It is necessary to make this variable independent and, in this case, only an arbitrary value can be equal to zero. For Jordan, exclusion with solving elements $\mathbf{R}_{kk}$ the equations will be written in the form

$$\left.\begin{aligned} \mathbf{S}_i^0 &= \sum_{j=1} \mathbf{R}_{ij}^* \mathbf{d}_j^0 + \mathbf{R}_{ik}^* \mathbf{S}_k^0 + \sum_{j=k+1} \mathbf{R}_{ij}^* \mathbf{d}_j^0 \qquad (i = 1, k-1) \\ \mathbf{d}_k^0 &= \sum_{j=1} \mathbf{R}_{kj}^* \mathbf{d}_j^0 + \mathbf{R}_{kk}^* \mathbf{S}_k^0 + \sum_{j=k+1} \mathbf{R}_{kj}^* \mathbf{d}_j^0 \end{aligned}\right\} \tag{11.8}$$

Elements in the new matrix $\mathbf{R}^*$ are marked in (11.8) with stars, the kth column and the kth row can be deleted from $\mathbf{R}^*$. This matrix is of the $(n–1)^{th}$ order and is also the projection matrix but for the structure without the kth element. If a new zero line will be discovered in this matrix, this fact implies that corresponding element of the system becomes absolutely necessary after the kth element had been removed. The process of removing can be continued until zero matrix would appear and the system would become statically determinate. One can state that the chain of removals brings to the loss of geometric invariability.

An indicator of the system's quality is often represented not only by the property of geometric invariability but also by dimensions of the remaining rigid body frame. This fact means that some failures are possible with the system's transformation into a mechanism but these failures have no global features. If the indicator is chosen by the relative number of joints incoming in the remaining rigid body frame, then values Y_i, Y_{ij}, and Y_{ijk} in (11.1) can be substituted with D_i, D_{ij}, and D_{ij}, respectively. These values characterize the joint's relative number when one, two, or three elements of the system are removed. This procedure determinates, for example, the complexity and the cost of structural repair. The use of D estimation instead of Y will give a new measure or index

$$\begin{aligned} \eta = \left(1 - \frac{1}{n}\right)^{n-1} \frac{1}{n} \sum_{i=1}^{n} D_i + \left(1 - \frac{1}{n}\right)^{n-2} \frac{1}{n^2} \sum_{j \le i}^{n} \sum_{i=1}^{n} D_{ij} \\ + \left(1 - \frac{1}{n}\right)^{n-3} \frac{1}{n^3} \sum_{j \le i}^{n} \sum_{i \le k}^{n} \sum_{k=1}^{n} D_{ij} + \ldots\ldots \end{aligned} \tag{11.9}$$

This index is also useful for structural estimation of statically determinate systems for which η_i was identically equal to zero. The number of terms in (11.9) will be more than $(r - 1)$. However, their feature of powerful damping remains intact and this fact makes it possible to keep in real calculations a rather small number of terms in the series. It is also possible to present the survivability indexes as follows:

$$\left.\begin{aligned} \xi_{yi} &= \eta_y - (\eta y,\, x_i = 0); \quad (i = 1,\ldots,n) \\ \xi_{di} &= \eta_d - (\eta d,\, x_i = 0); \end{aligned}\right\} \tag{11.10}$$

The removed members are determined in (11.10) using formula (11.9) but under the condition that in all considered versions the diagrams of the ith element will be included.

All these arrangements make it possible assessing the contribution of each element to the structural survivability and ranking the elements according to their importance.

To illustrate the above, six different trusses are schematically presented in Figure 11.1. Three of them, **a**, **c**, and **e**, are statically determinate and other three, statically indeterminate. All trusses have three extra supporting joints and seven elements. Some description data of the trusses are given in Table 11.1.

Sums ΣD and $\Sigma\Sigma D$ in Table 11.1 depend on the number of simultaneously removed elements. This situation is defined for all possible versions for the removed elements, if these elements are not indicated as absolutely reliable. Survivability indices η and ξ are calculated according to formula (11.10) and the results are

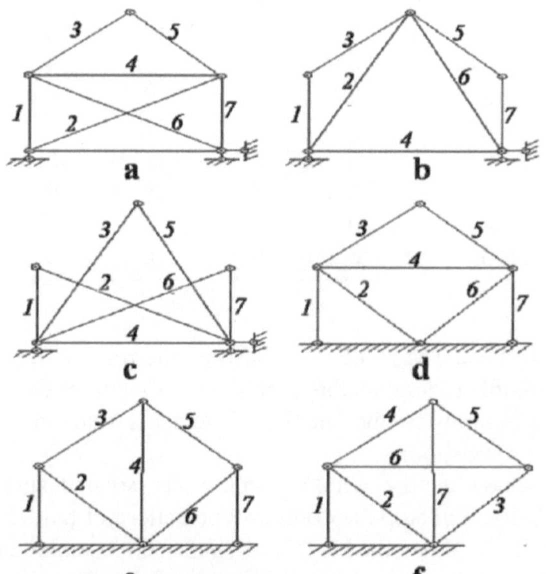

FIGURE 11.1 Schematics of trusses: a, c, & e—statically determinate; b, d, & f—statically indeterminate

TABLE 11.1
Characteristics of the trusses in Figure 11.1

Reliable elements	# of removed elements	Total number of joints in each truss					
		a	b	c	d	e	f
No absolutely reliable elements	1	4	8	12	19	21	18
	2	2	8	18	26	26	16
	3	0	4	20	24	22	12
	4	0	1	15	10	10	5
	5	0	0	0	2	2	1
Element # 1	1	4	6	10	16	18	18
$x_1 = 0$	2	2	4	12	21	20	16
	3	0	1	10	17	15	12
	4	0	0	7	6	8	5
	5	0	0	0	1	1	1
Element # 2	1	4	8	10	16	18	18
$x_2 = 0$	2	2	8	12	21	20	16
	3	0	4	10	17	15	12
	4	0	1	7	6	6	5
	5	0	0	0	1	1	1
Element # 3	1	2	6	10	17	18	15
$x_3 = 0$	2	0	4	12	14	16	10
	3	0	1	10	10	9	5
	4	0	0	7	2	2	1
	5	0	0	0	0	0	0
Element # 4	1	4	8	12	18	18	15
$x_4 = 0$	2	2	8	18	18	18	10
	3	0	4	20	10	8	5
	4	0	1	3	2	2	1
	5	0	0	0	0	0	0

presented in Table 11.2. The results demonstrate the role of absolutely necessary elements. It is possible to note that assembling the system of element in accord to the principles of autonomy for the function of elements (diagram "**c**" in Table 11.2) would improve its survivability.

The idea to subdivide the whole system into isolated subsystems is well known. This isolation will stop the domino-type pattern of failures. It was already mentioned in the Bible. When the Noah's Ark had been assembled God pointed out to Noah: "Subdivide Ark, and make tarring inside and outside …. and build it from the wood Gofer and Nimotriclin."

If the system has no absolutely necessary elements then every variable in the first sum in (10.1) will be equal to the unity. This situation can be summarized and in all other sums all items will not disappear and each of these sums will get the

TABLE 11.2
Survivability indexes η and ξ for different schemes of the truss

	Numerical values of indexes for trusses in Figure 11.1					
Estimate	a	b	c	d	e	f
η_d	0.082	0,178	0.295	0.453	0.491	0.397
ξ_{d1}	0	0.152	0.063	0.076	0.079	0
ξ_{d2}	0	0	0.063	0.076	0.079	0.057
ξ_{d3}	0.044	0.052	0.063	0.090	0.096	0.060
ξ_{d4}	0	0	0.001	0.089	0.084	0.080

maximum possible value. In that case $\Sigma Yi = n$ and it can supposedly be a model of survivability. If so, then:

$$\sum_{i=1}^{n} Yi = n; \quad \sum_{i \le j}^{n} \sum_{j=1}^{n} Yij = n(n-1)/2; \quad \sum_{j \le i}^{n} \sum_{i \le k}^{n} \sum_{k=1}^{n} Yijk = n(n-1)(n-2)/6; \; .... \tag{11.11}$$

Then instead of (11.1) a new expression for survivability index can be written

$$\eta_y^* = \sum_{i=1}^{r} \frac{(1-1/n)^{n-i}(1/n)^{in}!}{i!(n-i)!} = Bi(r, n, 1/n) - (1-1/n) \tag{11.12}$$

where

$$Bi(\text{run}\,1/n)$$

is the binominal distribution function.

Values of standard indexes η_y^* are given in Table 11.3 as a function of the number of elements in the system n and degree of static indeterminacy r. Their dependence on the number of elements, n, is rather weak. For most structures with the number of elements more than 25 it is possible to take an approximate estimate based on $n = 100$. Dependence of η_y^* on degree of static indeterminacy r is more pronounced but with r increasing its effect on the values of η_y^* decreases. Therefore greater values of r need not be considered.

Another method of approaching the discussed problem consists of introducing the survivability index η in the following form:

$$\eta = \frac{P_f}{P_f'} \tag{11.13}$$

where P_f is the probability of failure of the designed system; and P_f' is the probability of failure of the same system when some members failed. Survivability

TABLE 11.3
Standard index η_y^* vs. number of elements and degree of static indeterminacy

Degree of static indeterminacy	Values of η_y^* for different numbers of elements n				
	$n = 10$	$n = 25$	$n = 50$	$n = 100$	$n = 1000$
$r = 1$	0.3871	0,3754	0.3716	0,3698	0.3681
$r = 2$	0.5710	0.5593	0.5555	0.5537	0.5520
$r = 3$	0.5936	0.6206	0.6188	0.6150	0.6133
$r = 4$	0.6089	0.6359	0.6321	0.6303	0.6286
$r = 5$	0.6120	0.6390	0.6352	0.6334	0.6317
$r = 6$	0.6125	0.6395	0.6357	0.6338	0.6322

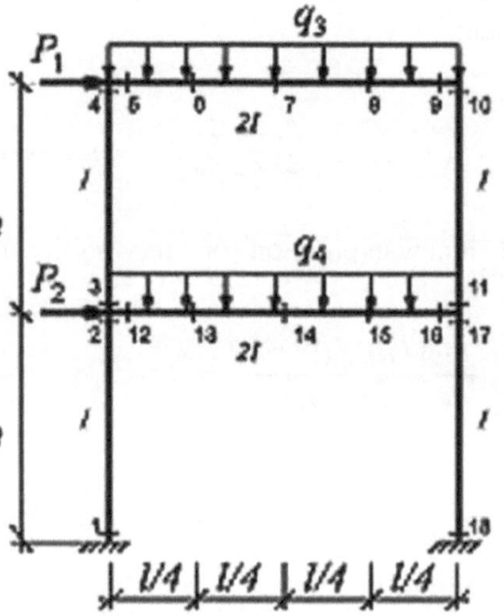

FIGURE 11.2 Two-story frame

factors η range within the interval of [0,1]. The greater the value of η, the greater is the reserve of survivability in the structural system. This is illustrated in the example of a steel frame shown in Figure 11.2.

The horizontal girders have a span of l = 6 m. The height of vertical columns h = 4 m. All members of the frame are made of I-beams with section modulus $W = 6.15 \cdot 10^{-5}$ m^3 for the 1st floor column, $W = 8.28 \cdot 10^{-5}$ m^3 for the 2nd floor column, $W = 1.270 \cdot 10^{-4}$ m^3 for the 1st floor girder, $W = 1.098 \cdot 10^{-4}$ m^3 for the 2nd floor girder.

Probabilistic analysis was performed considering the random nature of applied loads and yield stress of frame material, with the given distribution probabilities. Table 11.4 contains parameters of these distributions. Calculations were made using linear programming method (simplex method) using the direct integration of the distribution functions (Raizer, 2009). Probability of failure is $P_f = 5.51 \cdot 10^{-5}$.

TABLE 11.4
Probability distribution parameters for the frame

Random value	Distribution	Mean value	Standard deviation	Parameters of distribution	Design values
Wind load P_1, P_2	Gumbel	0.144 kPa	0.037 kPa	$u = 0.127$ kPa $z = 0.029$ kPa	0.2576 kPa
Snow load q_3	Gumbel	1.1418 kPa	0.4681 kPa	$u = 0.931$ kPa $z = 0.365$ kPa	1.6 kPa
Load due to use q_4	Gauss	0.88 kPa	0.21 kPa	–	1.68 kPa
Yield point σ_y	Weibul	305.25 MPa	25 MPa	$\beta = 14.3$; $x0 = 0$; $\alpha = 316.42$ MPa	245 MPa

TABLE 11.5
Probability of failure P_f at reduced section modulus *W* of different sections

	P_f at percentage points reduction of section modulus *W*					
Sect. #	5%	10%	25%	50%	75%	95%
1	5.51×10^{-5}	5.51×10^{-5}	5.51×10^{-5}	5.51×10^{-5}	7.53×10^{-5}	8.42×10^{-5}
2	5.51×10^{-5}	5.51×10^{-5}	5.51×10^{-5}	5.51×10^{-5}	7.41×10^{-5}	8.94×10^{-5}
3	5.51×10^{-5}	5.51×10^{-5}	5.51×10^{-5}	5.51×10^{-5}	5.51×10^{-5}	5.51×10^{-5}
4	5.83×10^{-5}	5.96×10^{-5}	0.000101	0.000207	0.000389	0.000570
5	5.51×10^{-5}	5.51×10^{-5}	8.42×10^{-5}	0.000122	0.000309	0.000547
6	5.51×10^{-5}	5.51×10^{-5}	5.51×10^{-5}	0.000107	0.000755	0.004562
7	6.19×10^{-5}	7.90×10^{-5}	0.000303	0.001246	0.006322	0.025580
8	5.51×10^{-5}	5.51×10^{-5}	5.51×10^{-5}	8.34×10^{-5}	0.000734	0.004771
9	5.51×10^{-5}	5.51×10^{-5}	5.51×10^{-5}	0.000137	0.000319	0.000593
10	5.95×10^{-5}	6.86×10^{-5}	0.000103	0.000207	0.000392	0.000564
11	5.51×10^{-5}	5.51×10^{-5}	5.51×10^{-5}	5.51×10^{-5}	5.51×10^{-5}	5.51×10^{-5}
12	5.51×10^{-5}	5.51×10^{-5}	5.51×10^{-5}	5.51×10^{-5}	0.000112	0.000265
13	5.51×10^{-5}	5.51×10^{-5}	5.51×10^{-5}	5.51×10^{-5}	0.000224	0.000873
14	5.51×10^{-5}	5.51×10^{-5}	5.51×10^{-5}	0.000890	0.001063	0.002327
15	5.51×10^{-5}	5.51×10^{-5}	5.51×10^{-5}	5.51×10^{-5}	0.000229	0.000871
16	5.51×10^{-5}	5.51×10^{-5}	5.51×10^{-5}	5.51×10^{-5}	0.000112	0.000259
17	5.51×10^{-5}	5.51×10^{-5}	5.51×10^{-5}	5.51×10^{-5}	7.30×10^{-5}	8.27×10^{-5}
18	5.51×10^{-5}	5.51×10^{-5}	5.51×10^{-5}	5.51×10^{-5}	7.34×10^{-5}	8.33×10^{-5}

More probable is the partial mechanism of failure when plastic hinges develop in the cross-sections 4, 7, and 9 (Figure 11.2). The failure probabilities of the frame are listed in Table 11.5 for different cases of weakening the cross-sections. Table 11.5 shows that if any cross-section fails, the failure probability for the frame

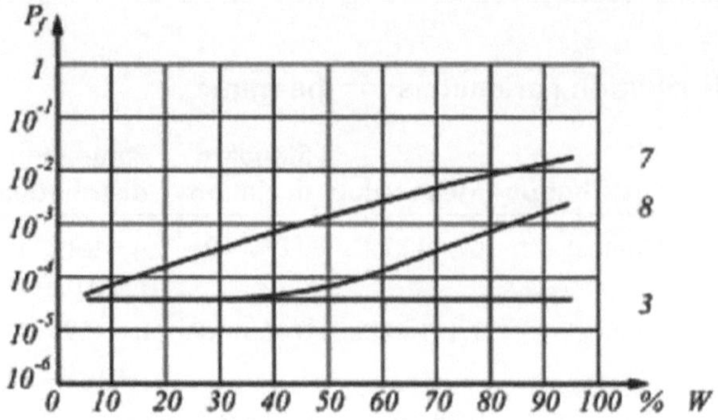

FIGURE 11.3 Probability of failure P_f of sections 3, 7 and 8 as a function of percentage reduction of their section modulus W

will not exceed the value of $P_f' = 0.0255$ (failure of cross-section 7). The failure of cross-section 7 will not lead to the collapse of the whole structure but essentially decreases its survivability. Even the full failure of cross-sections 2 or 11 has no influence on the probability of this frame. The failure of cross-sections 1, 2, 17, or 18 also has no essential effect on this probability. Survivability index of the frame with regard to the failure of cross-section 7 is $\eta = 0.02558$. If one cross-section failed, the failure probability for the whole frame will decrease to $P_f' = 0.00477$, and the survivability index will be: $\eta = 5.51*10^{-5}/0.00477 = 0.0115$. The failure probabilities for weakened cross-sections 7, 8, and 3 are plotted in Figure 11.3.

The above example, that can be considered as a test, demonstrates the importance of using during design both traditional deterministic and probabilistic methods. The latter make it possible to assess the influence of particular elements on the general reliability of the system and identify the elements to be strengthened.

11.5 STOCHASTIC ANALYSIS OF DYNAMIC INSTABILITY

In the analysis of nonlinear vibrations of structures, one of the most important issues is to study the behavior of the system after a long period of time t after the start of loading, that is, at $t \to \infty$ (Raizer, 1985, 2010). If the excitation force is represented as multiple periodic pulses, it causes the evolution of the nonlinear system in time. The study of the asymptotic behavior of nonlinear systems under the action of short-term periodic pulses is of interest. Consider the nonlinear parametric oscillations of a flexible rectangular plate, on the sides of which a and b are applied compressive impulsive loads N_x, N_y, following each other with a period of T.

When deriving the equations of motion of the plate, the Lagrange equations are used

$$\frac{\partial}{\partial t}\left(\frac{\partial L}{\partial \dot{q}_i}\right) - \frac{\partial L}{\partial q_i} = 0 \tag{11.14}$$

Where L = K-V is the Lagrange function; K is the kinetic energy of the system; V is the potential energy of the system; $q_i, \dot{q}_i$ make up the independent variables (generalized coordinates and velocities).

The kinetic energy is written as

$$K = \frac{1}{2}\frac{\gamma h}{g}\int_0^a\int_0^b\left(\frac{\partial w}{\partial t}\right)^2 dxdy \tag{11.15}$$

where $w = w(x, y)$ is the deflection function of the plate; h is the thickness of the plate; γ is the density of the material; and g is the acceleration of gravity.

The potential energy can be represented as

$$V = V_u + V_\partial + V_c \tag{11.16}$$

Here V_u is the potential bending energy of the plate;
V_∂ is the potential energy of deformation;
V_c is the potential of external forces.

The resolving equations for calculating the plate have the form (Volmir, 1967):

$$\begin{aligned} &\frac{D}{h}\nabla^2\nabla^2 w = L(w,\Phi) \\ &\frac{1}{E}\nabla^2\nabla^2\Phi = -\frac{1}{2}L(w,w) \end{aligned} \tag{11.17}$$

Here

$$D = Eh^3/12(1-\mu^2)$$

is the cylindrical stiffness; μ is Poisson's ratio; and w, F are the deflection and stress functions.

The differential operators in (11.17) have the form

$$\nabla^2\nabla^2\Phi = \frac{\partial^4\Phi}{\partial x^4} + 2\frac{\partial^4\Phi}{\partial x^2\partial y^2} + \frac{\partial^4\Phi}{\partial y^4}$$

$$\begin{aligned} &\nabla^2\nabla^2(w) = \frac{\partial^4 w}{\partial x^4} + 2\frac{\partial^4 w}{\partial x^2\partial y^2} + \frac{\partial^4 w}{\partial y^4} \\ &L(w,\Phi) = \frac{\partial^2 w}{\partial x^2}\frac{\partial^2\Phi}{\partial y^2} - 2\frac{\partial^2 w}{\partial x\partial y}\frac{\partial^2\Phi}{\partial x\partial y} + \frac{\partial^2 w}{\partial y^2}\frac{\partial^2\Phi}{\partial x^2} \\ &L(w,w) = 2\left[\frac{\partial^2 w}{\partial x^2}\frac{\partial^2 w}{\partial y^2} - \left(\frac{\partial^2 w}{\partial x\partial y}\right)^2\right] \end{aligned} \tag{11.18}$$

The plate is considered as a system with one degree of freedom, the curved surface of which is approximated by the equation

$$w = w_0(t)\sin\frac{\pi x}{a}\sin\frac{\pi y}{b} \tag{11.19}$$

The stress function is derived from the strain compatibility Equation (11.17):

$$\Phi(x,y) = E\frac{w_0}{32}\left[\left(\frac{a}{b}\right)^2\cos\frac{2\pi x}{a} + \left(\frac{b}{a}\right)^2\cos\frac{2\pi y}{b}\right] - \frac{N_x y^2}{2} - \frac{N_y x^2}{2} \tag{11.20}$$

It is assumed that the longitudinal edges of the plate are fixed. If Δ_x, Δ_y make up the mutual convergence of the edges, then $\Delta_y = 0$, and

$$N_y = \mu N_x - \frac{E\pi^2}{8}\left(\frac{w_0}{b}\right)^2 \tag{11.21}$$

By introducing a dimensionless deflection $q = \frac{w_0}{h}$, denoting $\lambda = \frac{a}{b}$, substituting (11.19) and (11.20) into (11.15) and (11.16), the expressions for kinetic and potential energy will take the form

$$\begin{aligned} K &= \frac{m}{2}\left(\frac{\partial q}{\partial t}\right)^2 \\ V &= \frac{C}{2}\left[\left(1 - \frac{N}{N_{kp}}\right)q^2 + \frac{\alpha q^2}{2}\right] + rN^2 \end{aligned} \tag{11.22}$$

where

$$m = \frac{ab\gamma h^3}{4q};\ \alpha = \frac{3(3\lambda^4+1)(1-\mu^2)}{4(1+\lambda^2)^2};\ C = \frac{\pi^4 Eh^3(1+\lambda^2)^2}{48(1-\mu^2)\lambda^2 ab};\ N = N_x\frac{b}{Eh};$$

$$N_{kp} = \frac{\pi^2(1+\lambda^2)^2}{12(1-\mu^2)(1+\mu\lambda^2)\lambda^2};\ r = \frac{Eh^3(1-\mu^2)\lambda}{2b^2}$$

By substituting (11.22) into (11.14), we obtain the nonlinear equation

$$\frac{d^{2q}}{dt^2} + \omega^2(1+\alpha q^2)q - \frac{N}{N_{kp}^2}\omega^2 q = 0 \tag{11.23}$$

where

$$\omega = \frac{\pi^2 h\sqrt{Eg}(1+\lambda^2)}{ab\lambda\sqrt{12\gamma(1-\mu^2)}}$$

This equation is similar in appearance to the one discussed in Volmir (1967), but the compressive periodic value of the load is assumed here in the form

$$N = N_t \sum_{n=-\infty}^{\infty} \delta(t - nT) \tag{11.24}$$

where

$$T = \frac{2\pi}{\Omega}$$

and Ω is the frequency of the applied load.

In the interval between the pulses, Equation (11.23) will take the form

$$\frac{d^2 q}{dt^2} + \omega^2 (1 + \alpha q^2) q = 0 \tag{11.25}$$

The solution of nonlinear Equation (11.25) can be obtained by the small parameter method

$$q = A\cos[(\omega + \Delta\omega)t + \phi]; \Delta\omega = \frac{3\alpha A^2 \omega}{8} \tag{11.26}$$

Essential for the parametric nonlinear system is the fact that for sufficiently large perturbation actions with a wide spectrum (including a large number of harmonics), the motion of the nonlinear system can acquire a random character. The phase change begins to be close to random. The energy of the system increases on average over time. The phenomenon of stochastic instability, considered in (Zaslavsky, 1984), arises in connection with the motion of elementary particles in accelerators. To study this dynamical system, it is advisable to proceed to the canonical Hamilton equations of motion (Goldstein, 2002). The Hamilton function of the system is defined by the formula

$$H = K + V = \frac{p^2}{2m} + \frac{C}{2}\left[\left(1 - \frac{N}{N_{kp}}\right) q^2 + \frac{\alpha q^4}{2}\right] + rN^2 \tag{11.27}$$

Here $p = \frac{\partial K}{\partial \dot{q}}$ is the generalized impulse.

For the repeatedly periodic action (11.24), it is advisable to turn to the variables applicable in statistical mechanics, the angular variable ψ and the momentum canonically conjugate to ψ, the variable of action I. They are related to the p and q with relations

$$q = \sqrt{\frac{2I}{m\omega}} \sin \psi; p = \sqrt{2 \operatorname{Im} \omega} \cos \psi \tag{11.28}$$

Using the perturbation method (Prigogine, 2017), the Hamilton function can be represented in the form

$$H = H_0 + \nu H_1 \tag{11.29}$$

Here

$$H_0 = \frac{p^2}{2m} + \frac{Cq^2}{2}$$

is the Hamiltonian of an unperturbed system (one-dimensional harmonic oscillator); ν is a small parameter; C is a coefficient associated with the frequency ω by the ratio $\omega = (C / m)^{1/2}$; and νH_1 denotes a perturbing function. For unperturbed motion, $I = \text{const}$, and ψ depends linearly on time, and at the same time

$$H_0 = \omega I.$$

For perturbed motion, the action-phase variables satisfy a system of differential equations:

$$\begin{aligned} \frac{\partial \psi}{\partial t} &= \frac{\partial H_9}{\partial I} + \nu \frac{\partial H_1}{\partial I} \\ \frac{\partial I}{\partial t} &= -\frac{\partial H_0}{\partial \psi} - \nu \frac{\partial H_1}{\partial \psi} \end{aligned} \tag{11.30}$$

Substituting (1127) in (11.30), taking into account (11.29), the canonical equations of the first approximation take the form

$$\begin{aligned} \frac{\partial \psi}{\partial t} &= \omega I \\ \frac{\partial I}{\partial t} &= \varepsilon I \sin 2\psi \sum_{-\infty}^{\infty} \delta(t - nT) \end{aligned} \tag{11.31}$$

where

$$\varepsilon = \frac{N\omega}{N_{kp}}$$

It is assumed that the angular variable and the action variable found from the consideration of the unperturbed motion are taken as canonical variables for the perturbed motion.

It follows from Equation (10.31) that in the interval between two pulses, $I = \text{const}$.

As a result, from the nth pulse, the action variable abruptly changes by the value:

$$\Delta I = \varepsilon I_n \sin 2\psi_n \tag{11.32}$$

where I_n is the value of I at a given time

$$t = t_n.$$

The frequency is reduced by the value of

$$\Delta\omega = \frac{d\omega}{dI}\Delta I = \varepsilon I_n \sin 2\psi_n \frac{d\psi}{dI} \tag{11.33}$$

Hence, the phase will change as

$$\psi_{n+1} = \psi_n + V_n \sin 2\psi_n \tag{11.34}$$

where

$$V_n = \frac{\varepsilon I_n}{\Omega} \frac{d\omega(I_n)}{dI_n}$$

Equation (11.34) was considered in work by Zaslavsky (1984). The results of the numerical study showed that for $V_n > 1$ the motion will have an unordered stochastic character. To do this, you need to evaluate the normalized function:

$$R = \frac{\int_0^1 (\psi_{n+1} - \overline{\psi}_{n+1})(\psi_n - \overline{\psi}_n)\, d\psi_n}{\int_0^1 (\psi_n - \overline{\psi}_n)^2\, d\psi_n} \tag{11.35}$$

This function should be evaluated for large values of R using the transformation (10.34). Such an estimate has the form $J \approx 1/V$. Then the condition

$$V = \varepsilon\Delta\omega/\Omega > 1$$

holds. Or taking into account (11.26) we get:

$$\frac{3\varepsilon\alpha A^2\omega}{8\Omega} > 1, \text{п}$$

for

$$\Omega << \omega \tag{11.36}$$

Equation (10.36) can be considered as a condition when the phases behave chaotically, with a sufficiently large T, where the correlation between successive phases ψ_n and ψ_{n+1} is lost, and the process becomes approximately Markov and stochastic instability can develop in the system. To describe the behavior of the system, we can use the Fokker-Planck-Kolmogorov kinetic equation. If there are delta functions in the right part (11.31), it is convenient to proceed from the Liouville equation:

$$\frac{\partial f}{\partial t} + \frac{\partial}{\partial \psi}(\dot{\psi} f) + \frac{\partial}{\partial I}(\dot{I} f) = 0 \tag{11.37}$$

Here $f(I, \psi/t)$ denotes the density of the joint distribution of the angular variable and the action. Substituting (11.30) into (11.37), the Liouville equation takes the form

$$\begin{gathered} \frac{\partial f}{\partial t} + \omega(I)\frac{\partial f}{\partial \psi} = -\frac{\partial}{\partial I}(uf) \\ u = I\sin 2\psi \sum_{n=-\infty}^{\infty} \delta(t - nT) \end{gathered} \tag{11.38}$$

Given that $f(I, \psi/t)$ should be a periodic function ψ, we decompose it into a Fourier series in $e^{in\psi}$ (similar to u). For the amplitudes, the equation is obtained:

$$\frac{\partial f_n}{\partial t} + in\omega(I) f_n = -\varepsilon\frac{\partial}{\partial I}\sum_k (u_k f_{n-k} + u_{-k} f_{n+k}) \tag{11.39}$$

The problem is to construct an asymptotic solution that is valid when $t \to \infty$. Equation (11.39) is solved by iterations over ε. In this case, only the first term f_0 remains in the expansion of f, since this value represents the energy distribution density. In the problem under consideration, we can use the initial conditions of the form $f_n, t = 0$ for all $n \neq 0$, which are equivalent to the approximation of chaotic phases at the initial moment of time. Then, limiting in (10.39) of the terms to ε^2 and keeping among them only the main ones (giving the maximum contribution at $t \to \infty$.) and passing to the t-representation, we write

$$\frac{\partial f_0}{\partial t} = 8\pi\varepsilon^2\Omega \frac{\partial}{\partial I}\left[I \frac{\partial}{\partial I}\left(If_0\right)\right] \tag{11.40}$$

Equation (11.40) can be solved for given boundary and initial conditions. If we limit ourselves to a qualitative result, then multiplying Equation (10.40) by I and integrating, we find

$$\frac{d\bar{I}}{dt} = \theta\bar{I} \tag{11.41}$$

Here

$$\theta = 8\pi\varepsilon^2\Omega; \bar{I}$$

denotes the mathematical expectation of the action.

For $t = 0$

$\bar{I} = I_0$

Then the solution of Equation (11.41) becomes

$$\bar{I} = I_0 \exp(\theta t) \tag{11.42}$$

where $\bar{I} = I_0$ is the boundary beyond which randomness cannot occur, and the boundary corresponds to equality:

$$\frac{3\alpha\varepsilon A^2\omega}{8\Omega} = 1$$

In the unperturbed motion $q = A\cos(\omega t + \varphi)$. Then the Hamiltonian of unperturbed motion is equal to

$$H_0 = \omega I_0 = 0,5A^2C$$

Hence $\omega A^2 = \dfrac{2I_0}{m}$.

Consequently,

$$I_0 = \frac{4m\Omega}{3\alpha\varepsilon}$$

and the expression for the mathematical expectation of the action (11.42) will take the form

$$\bar{I} = \frac{4m\Omega}{3\alpha\varepsilon}\exp(8\pi\varepsilon^2\Omega t) \tag{11.43}$$

To solve the problem in the first approximation, we need to substitute (11.43) into the Hamiltonian:

$$\frac{m\dot{q}}{2} + \frac{Cq^2}{2} \approx \bar{I}\,\omega \tag{11.44}$$

Equation (11.44) is converted to the form

$$\dot{q} = \omega\left(\frac{2\bar{I}}{m\omega} - q^2\right)^{1/2} \tag{11.45}$$

the solution of which is

$$q = \sqrt{\frac{8\Omega}{3\varepsilon\alpha\omega}}\exp\left(4\pi\varepsilon^2\Omega t\right)\sin(\omega t + \varphi) \tag{11.46}$$

The exponential "divergence" of the amplitude is obvious from Equation (11.46). Energy is constantly poured into the system and a stager-peaked mode is implemented. The applied action contains a sufficiently large number of harmonics, and the influence of each harmonic is so great that the system is not able to stay close to any one resonance. There is a peculiar wandering of the system along the resonances, in which there are no integrals of motion, except for trivial ones.

A statistical approach to the analysis of the dynamic instability of a deterministic nonlinear system under a deterministic action in the form of repeated short-term pulses allows us to estimate the average displacement of the system.

11.6 EFFECTS OF UNEVEN FOOTING SETTLEMENT

Let us consider how the uneven precipitation of the base, which changes over time due to consolidation processes in the soil, affects the reliability of the structure (Raizer, 1986). The failure of the structure may occur before the sediment reaches the stabilizing value, due to a violation of the conditions of normal operation. The one-dimensional problem of soil compaction is considered. The determination of the hydrodynamic stress in the soil skeleton is reduced to the integration of the equation (Florin, 1951):

$$\frac{d^2\sigma_z}{dt^2} = c^2 \frac{d^2\sigma_z}{dz^2} \tag{11.47}$$

Where t is time; z is the coordinate for the depth of the ground;

$$c^2 = k(1+\varepsilon_{cp})\gamma_0 a$$

is the coefficient of consolidation of soil; k is the filtration coefficient; ε_{cp} denotes the average porosity coefficient; γ_0 is the density of water; and a is the compaction coefficient.

The pronounced heterogeneity of the soils is reflected in the variability of the characteristics, including the coefficient of consolidation, which should be considered as a random function of t and z. Here we propose an approximate method based on the assumption that the characteristics of the soil are random variables. Equation (10.47) is solved by the Fourier method under the following boundary and initial conditions: $t = 0$, $0 \le z \le h$, $\sigma_z = 0$ (h is the layer thickness); $t > 0$, $z = h$, $\sigma_z = F(t)$. The expression for the precipitation of the soil layer is obtained in the form

$$y_n(t) = \frac{8k}{\gamma_0 h}\int_0^t F(\xi)m = \sum_{m=1}^{\infty} \exp\left[-\lambda_m^2(t-\xi)\right]d\xi \tag{11.48}$$

from the action of a time-varying load $F(t)$ applied to the ground surface.

As a design scheme of the structure, you can take a continuous beam with time-shifting supports. The equation of the three moments when considering the settings of the supports for the same l spans of constant stiffness EI will have the form

$$M_{n-1}(t) + 4M_n(t) + M_{n+1}(t) = -\frac{6R_n^{\Phi}}{l} - \frac{6EI}{l^2}\left[y_{n+1}(t) - 2y_n(t) + y_{n-1}(t)\right] \tag{11.49}$$

Here R^{Φ} completes the dummy reaction of the nth support.

The sediment of the base $y_n(t)$ of the nth support, located on a pliable ground, is determined by the formula (10.48), where

$$F(t) = R_n / A_n; R_n(t)$$

is the value of the reactive response on this support at time t; A_n is the area of the base of the nth support. At the same time, the reaction of the nth support is expressed in terms of the support moments of the adjacent spans and the reaction of the nth support in the main system. The ratio (10.49) is reduced to the form

$$M_{n-1}(t)+4M_n(t)+M_{n+1}(t)=\eta(t)+$$
$$a_1\int_0^t\left[M_{n+2}(\xi)-4M_{n+1}(\xi)+6M_n(\xi)-4M_{n-1}(\xi)+M_{n-2}(\xi)\right]k(t-\xi)d\xi \quad 11.50)$$

where

$$\eta_n(t)=-\frac{6R_n^{\Phi}}{l}-\frac{6EIkh}{\gamma_0 Al^2c^2}\left(R_{n+1}^0-2R_n^0+R_{n-1}^0\right)\left(1-\frac{8}{\pi^2}\sum_{m=1,3,5}^{\infty}\frac{1}{m^2}\exp(-\lambda^2 mt)\right);$$
$$k(t-\xi)=\sum_{m=1,3,5}\exp\left[-\lambda_m^2(t-\xi)\right];$$
$$a_1=-\frac{48EIk}{\gamma_0 hAl^3}.$$

The solution of the integral equations of the five moments (10.50) can be obtained using the one-way Laplace transform (Arutjunjan, 1952).

For example, consider a two-span beam (Figure 11.4).

It is assumed that at some point t, conventionally taken as the initial item, a uniformly distributed load $F(t)$ was applied to this beam and the setting of the middle support began.

The supporting moment is determined by the solution of Equation (11.50):

$$M_{cp}(t)=-\frac{\beta_1+\beta_2}{1+b}+2(\beta_2-b\beta_1)\sum_{m=1,3,5}^{\infty}\left(\frac{1}{b+\cos^2\lambda_m}-\frac{1}{b}\right)\exp\left(-\frac{\lambda_m^2 t}{a_0^2}\right) \quad (11.51)$$

where

$$b=\frac{\beta h^2}{8c^2};\ \beta=\frac{36EI}{\gamma_0 Ahl^3}$$
$$\beta_1=M_{cp}(0)=-\frac{Fl^2}{8}$$
$$\beta_2-\frac{3EIkhR_2^0}{\gamma_0 Ac^2l^2};\ a_0=\frac{h}{2c}$$

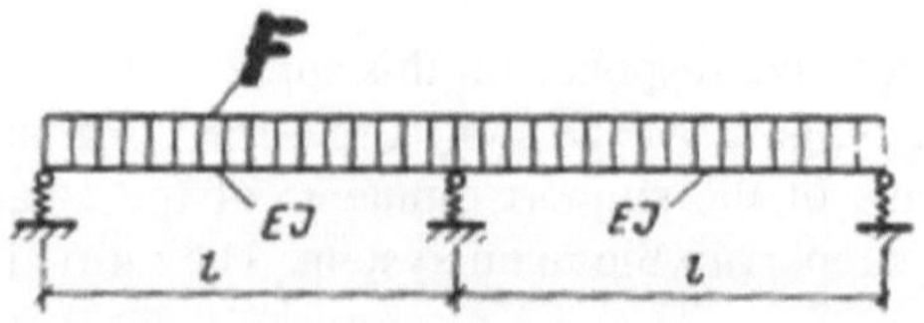

FIGURE 11.4 A design scheme

and λ_m denotes the roots of the transcendental equation:

$$tg\lambda = -\frac{\lambda}{b}$$

For

$$T = 0 \ M_{cp} = -\frac{Fl^2}{8}$$

for

$$t \to \infty \, M_{cp} = \frac{\beta + \beta_2}{1+b}$$

The case when the load is applied instantly is considered, for example, the setting of the soil layer as expressed by the formula

$$y_n(t) = \frac{8kR_n(0)h}{\gamma_0 A\pi^2 c^2}\left[\frac{\pi^2}{8} - \sum_{m=1,3,5}^{\infty} \frac{1}{m^2}\exp\left(-\lambda_m^2 t\right)\right] \tag{11.52}$$

where $R_n(0)$ denotes reaction in support.

For $t = 0$ $y = 0$, for $t \to \infty$ $y_n(t) = \dfrac{R_n(0)kh}{\gamma_0 Ac^2}$.

Let the area of non-failure work be determined by the bending moment on the middle support. If we constrain to only with one member of the series in (10.51) and make the logarithm of both parts of the equality, we get

$$W = log\ U = rt \tag{11.53}$$

Here it is indicated that

$$U = \frac{M_{cp} - X}{Y}; \quad X = \frac{\beta_1 + \beta_2}{1+b}; \quad Y = \frac{2(b\beta_1 - \beta_2)\cos^2\lambda_1}{b(b+\cos^2\lambda_1)}; \quad r = -\frac{\lambda_1^2}{a_0^2}$$

Let the area of no-failure work be determined by the deflection of the middle support. For the example in question, we have

$$R_{cp}(0) = 1,25Fl.$$

Assuming $m = 1$, we determine

$$\ln\left(\frac{\pi^2}{8} - \frac{\pi^2\gamma_0 Fc^2}{10kh}y\right) = rt \tag{11.54}$$

where

$$r = -\frac{\pi^2 c^2}{h^2}$$

Denoting

$$\frac{\pi^2}{8} - \frac{\pi^2 \gamma_0 F c^2}{10kh} y = U$$

Equation (10.54) is also converted to Equation (10.53). However, here the random function of time is not the bending moment, but the deflection. The boundary of the fault-free area is characterized by a random variable

$$U = U_0(W = W_0)$$

having a mathematical expectation and variance equal to

$$\bar{W}_0, s^2(W_0)$$

respectively.

Assuming that the failure occurs when

$$M_{cp} = M_{cp}^0$$

or when

$$y_{cp} = y_{cp}^0$$

then the probabilistic properties $U_0(W_0)$ will be determined only by the probabilistic properties of M_{cp}^0 or y_{cp}^0 in Equation (11.53), and the probabilistic properties of r are determined by the variability of the consolidation coefficient. The moment of failure occurrence corresponds to the intersection point of the implementation of the random time function (11.53) with the possible value of the boundary of the no-failure works (Figure 11.5).

At the time of the failure $r = W_0 / T$. it is assumed that r and W_0 are independent random variables and have a normal distribution truncated in the interval $(0, \infty)$.

This is possible, in particular, if the random variable U has a logarithmically normal distribution. The uptime distribution function will have the form

$$P(t) = a_{W_0} a_r \int_0^{\infty} \left[\int_{W_9/T}^{\infty} p(r) dr \right] p(W_0) dW_0 \qquad (11.55)$$

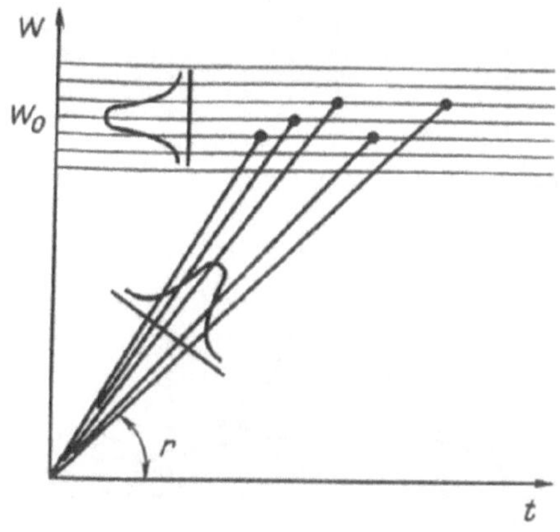

FIGURE 11.5 Versions of a random function

where a_{W_0}, a_r the normalizing factors of this distribution and

$$a_{W_0} = \left[\frac{1}{2} + \Phi\left(\frac{\bar{W}_0}{s(W_0)}\right)\right]^{-1} \quad a_r = \left[\frac{1}{2}\Phi\left(\frac{\bar{r}}{s(r)}\right)\right]^{-1}$$

To fulfill before the conversion (Druzhinin, 1977):

$$\frac{d}{dt}\int_{W_0/T}^{\infty} p(r)dr = \frac{W_0}{s(r)t^2\sqrt{2\pi}}\exp\left[-\frac{1}{2}\left(\frac{W_0}{s(r)t} - \frac{\bar{r}}{s(r)}\right)^2\right] \tag{11.56}$$

Differentiating (11.55) and considering (11.56), we write an expression for the uptime distribution density:

$$p(t) = \frac{a_{W_0} a_r \delta \exp\left[-0,5\left(\chi^2 / \delta\right) + \alpha^2\right]}{2\pi\left(t^2 + \delta^2\right)} + \frac{a_{W_0} a_r \left(\chi t + \alpha\delta^2\right)}{\sqrt{2\pi}\left(t^2 + \delta^2\right)}\exp\left[-\frac{(\chi - \alpha t)^2}{2\left(t^2 + \delta^2\right)}\right]\left[\frac{1}{2} + \Phi\left(\frac{\chi t + \alpha\delta^2}{\delta\sqrt{t^2 + \delta^2}}\right)\right] \tag{11.57}$$

where

$$\delta = \frac{s(W_0)}{s(r)}; \chi = \frac{\bar{W}_0}{s(r)}; \alpha = \frac{\bar{r}}{s(r)}$$

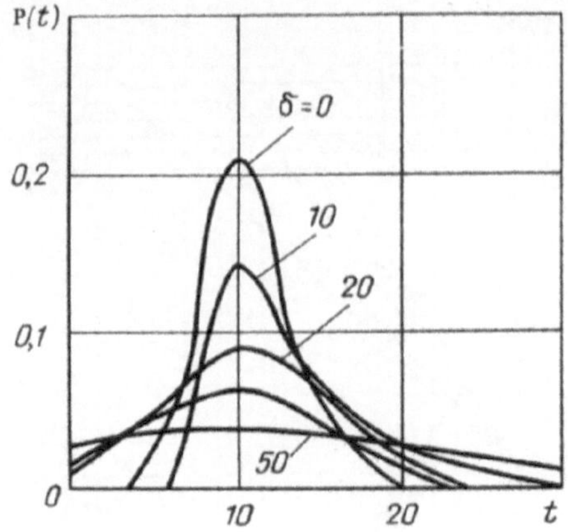

FIGURE 11.6 Uptime distribution density

For example, let the coefficient of variation of the consolidation coefficient be 0.2, and $\chi = 50$. Then, for different values of δ in accordance with (11.57), different density curves of the uptime distribution are obtained in Figure 11.6.

It follows from Figure 11.6 that the density of the uptime distribution increases, starting from a fixed value $t = t_i$, and then decreases with growth of t.

In a more correct formulation, the problem of constructing the reliability function of a structure with uneven foundation precipitation should be solved, if the necessary statistical data are available, considering the accumulation of uneven sediments as a random process over time. The approach presented here allows us to approximate the reliability of structures with uneven precipitation of the bases.

12 Conclusion

I can live with doubt, and uncertainty, and not knowing. I think it's much more interesting to live not knowing than to have answers which might be wrong. I have approximate answers, and possible beliefs, and different degrees of certainty about different things, but I'm not absolutely sure of anything, and in many things, I don't know anything about, such as whether it means anything to ask why we're here, and what the question might mean. I might think about a little, but if I can't figure it out, then I go to something else. But I don't have to know an answer. I don't feel frightened by not knowing things, by being lost in a mysterious universe without having any purpose, which is the way it really is, as far as I can tell, possibly. It doesn't frighten me.

Richard Feynman, *The Pleasure of Finding Things Out: The Best Short Works of Richard P. Feynman*

Maturity is the capacity to endure uncertainty.

John Finley

We must become more comfortable with probability and uncertainty.

Nate Silver

Ich bin kein ausgeklügelt Buch,
Ich bin ein Mensch mit seinemWiderpruch.
(I am not a contrived book, but a human being with all its contradictions).

C. F. Mayer

This book presented three competing philosophies of modeling uncertainty: Probabilistic concept of reliability, safety-factor based on fuzzy sets, and convex modelling. The first two possess a measure: The first one deals with probability density functions, whereas fuzzy sets-based analysis incorporates the notion of the membership function. The convex modelling, on the other hand, does not invoke the measure. It deals only with constraints which characterize the bounded uncertainty. The paramount question arises, which one ought to be embraced by the designer? This reminds us of the short poem of John Godfrey Saxe (1816–1887), titled "The Blind Men and the Elephant". It appears instructive to reproduce it fully:

It was six men of Indostan/To learning much inclined,
Who went to see the Elephant/(Though all of them were blind),
That each by observation/Might satisfy his mind.

DOI: 10.1201/9781003265993-12

The First approached the Elephant,/And happening to fall
Against his broad and sturdy side,/At once began to bawl:
"God bless me! but the Elephant/Is very like a WALL!"

The Second, feeling of the tusk,/Cried, "Ho, what have we here,
So very round and smooth and sharp?/To me 'tis mighty clear
This wonder of an Elephant Is very like a SPEAR!"

The Third approached the animal,/And happening to take
The squirming trunk within his hands,
Thus boldly up and spake:
"I see," quoth he, "the Elephant/Is very like a SNAKE!"

The Fourth reached out an eager hand,/And felt about the knee
"What most this wondrous beast is like/
Is mighty plain, "quoth he:
"'Tis clear enough the Elephant Is very like a TREE!"

The Fifth, who chanced to touch the ear,/Said: "E'en the blindest man
Can tell what this resembles most;/Deny the fact who can,
This marvel of an Elephant Is very like a FAN!"

The Sixth no sooner had begun/About the beast to grope,
Then seizing on the swinging tail/That fell within his scope,
"I see," quoth he, "the Elephant/Is very like a ROPE!"

And so these men of Indostan/Exceeding stiff and strong,
Though each was partly in the right,/And all were in the wrong!
Saxe then delivers the moral of the story:

So oft in theologic wars,/The disputants, I ween,
Rail on in utter ignorance/Of what each other mean,
And prate about an Elephant/Not one of them has seen!

Wikipedia (2021) informs that "The earliest versions of the parable of blind men and elephant is found in Buddhist, Hindu and Jain texts, as they discuss the limits of perception and the importance of complete context. The parable has several Indian variations." Moreover,

"In some versions, the blind men then discover their disagreements, suspect the others to be not telling the truth and come to blows. The stories also differ primarily in how the elephant's body parts are described, how violent the conflict becomes and how (or if) the conflict among the men and their perspectives is resolved. In some versions, they stop talking, start listening and collaborate to "see" the full elephant. In another, a sighted man enters

FIGURE 12.1 Six blind men and elephant (courtesy of Wikimedia Commons)

the parable and describes the entire elephant from various perspectives, the blind men then learn that they were all partially correct and partially wrong."

Remarkably, in the definitive paper by Elizabeth Paté-Cornell (1996) there are "six levels of treatment" of uncertainties in risk analysis.

As Werner Heisenberg (1958) stated in his book Physics *and Philosophy: The Revolution in Modern Science*, "We have to remember that what we observe is not nature in itself, but nature exposed to our method of questioning."

Thus, to reiterate, this book presented three competing philosophies of modelling uncertainty: Probabilistic concept of reliability, safety-factor based on fuzzy sets, and convex modelling. The first two possess measure: The first one deals with probability density functions, fuzzy-sets based analysis incorporates the notion of the membership function, whereas the convex modelling does not invoke the measure. It sems only with constraints which characterize the bounded uncertainty. The paramount questiona rises which one ought to be embraced by the designer?

Jennings (2011) advances the following ideas: "In many ways this poem [of Saxe] is an excellent metaphor for science and demonstrates how science works. The blind men want to learn about the elephant, so they make observations. Good so far – in science, observations rule. They then make models to describe what they have seen, or in this case, felt. Again, good scientific methodology: the tusk is indeed well modeled as a spear. Since each has chanced upon a different part of the elephant, the models differ. This not a problem. In science we always model different parts of reality differently; we do not use the standard model of particle physics to describe planetary orbits or cell division.

But the blind men then make the classic error and assume theirs is the global model that describes the entire elephant. The scientific method would have required them to make predictions based on their models and test them against additional observations. This would have revealed the problem – and the elephant. Instead, they argued. This, of course, bears no resemblance to real scientists who never argue heatedly on the basis of partial data. Nope, never, has not happened (OK, stop thinking of your colleague).

The moral is also interesting (not just for theologic war but also scientific ones): *Rail on in utter ignorance of what each other mean*. This seems like a foreshadowing of Kuhn's incommensurability of paradigms. Their ideas and frames of reference are so different they cannot understand each other.

And prate about an Elephant Not one of them has seen! Here we come to the nub. To a large extent science is precisely this, prating about an elephant we have not seen. We can learn a lot about the elephant we have not seen by making repeated observations of the beast and learning how the various models fit together and interlock with each other. We may never, by feel or even by sight, be able to construct the ultimate model of the elephant but we can obtain a lot of useful information and construct quite accurate models of at least some aspects of the elephant. The trick is not to make the mistake of assuming one's partial model is the whole truth. This mistake has been made repeatedly: Newton's laws of motion, Maxwell's equation, the fixed continents … There is now a talk of a theory of everything (TOE); the same mistake being made again.

The blind men's error has been common in the philosophy of science as well. Different models for the scientific method capture different aspects of the problem: induction, hypothesis, logical positivism (verification), paradigms, falsification, and even no method. The philosophers do indeed resemble the blind men disputing loud and long. The real task is to see how the various models fit together to give a coherent whole. Like the models of the elephant, the different approaches to the scientific method are not so much wrong as incomplete. The approach I advocate of models competing against each other in making successful predictions is an attempt at a more unified approach; one can see how the various precursor models are aspects of it. But as always, this will probably be just one aspect of a more complete approach that can never be completely known and will always be debated. But for most practical purposes, the scientific method can be, and probably is, known well enough. Blind men, even those with a philosophical bent, can learn a lot about an elephant by groping in the dark.

Mahmoudi (2020) writes:

> Each man provides accurate, but only partial, information, completely different from the other men. They cannot synthesize an accurate picture of an elephant that encompasses their distinct experiences of the trunk, the ear, the leg, the side, the tail and a tusk. In other words, working exclusively in our own siloed fields of study constricts what we can see and achieve.

Instead, combining our different individual lenses/experiences/expertise in a constructive and receptive discussion platform may enable the scientific community to solve urgent and/or important problems in a more effective and timely manner.

Jennings' (2015) book has a chapter entitled "Philosophy Is Not Always Useless". So, which philosophy do we prefer, probabilistic, fuzzy states-based thinking, or convex modelling? Here we quote the American novelist, essayist, screenwriter, and short story writer F. Scott Fitzgerald (1896–1940): "The test of a first-rate intelligence is the ability to hold two opposed ideas in the mind at the same time and still retain the ability to function." In view of the above, we ought to subscribe to each and every uncertainty analysis, depending on circumstances and availability and the character of the data.

Each of the analyses has its own advantages and drawbacks. These drawbacks ought not be overlooked as people are not noticing the 800-pound elephant in the room, the latter being a metaphor for unnecessary neglect.

Each of the uncertainty modeling techniques was subjected to criticism, sometimes even harsh one. According to Bart Kosko (1993), "Probability has turned modem science into a truth casino." Elsewhere, he claims (Kosko, 1994):

> Probability does not exist. The authors Laviolette and Seaman (1994) grant this at the start and tum to deFinetti (1974) for support. I agree that probability does not exist in the physical sense. It does not take up space or time in the space-time continuum we call our world. The same holds true for the existence of numbers.

He adds emphasis: "The probability monopoly is over." Not all would agree with such a strong statement. According to Pierre Simon, Marquis de Laplace

FIGURE 12.2 The 800-pound elephant in the room

(1749–1827) said as much before Kosko: "The most important questions of life are, for the most part, really only problems of probability." Klir (1994) states in more tolerant manner: "I argue in the paper that probability theory is capable of representing only one type of uncertainty; to capture the full scope of uncertainty, I argue, one has to go beyond probability theory."

Now, let us discuss opposition to fuzzy sets-based design. Rudolf Kalman (1930–2016) (see his paper of 1972) stated (Zadeh, 2004):

> Let me say quite categorically that there is no such thing as a fuzzy concept … We do talk about fuzzy things, but they are not scientific concepts. Some people in the past have discovered certain interesting things, formulated their findings in a non-fuzzy way, and therefore we have progressed in science.

Elsewhere, Kalman (1930–2016) (see Zadeh, 2011) maintained:

> I would like to comment briefly on Professor Zadeh's presentation. His proposals could be severely, ferociously, even brutally criticized from a technical point of view. This would be out of place here. But a blunt question remains: Is professor Zadeh presenting important ideas or is he indulging in wishful thinking? No doubt Professor Zadeh's enthusiasm for fuzziness has been reinforced by the prevailing climate in the U. S.-one of unprecedented permissiveness. Fuzzification, is a kind of scientific permissiveness; it tends to result in socially appealing slogans unaccompanied by the discipline of hard scientific work and patient observation.

However, Zadeh (2015) maintained: "My crystal ball is fuzzy." Additionally, he stated

> A fuzzy set is a class of objects with a continuum of grades of membership. Such a set is characterized by a membership (characteristic) function which assigns to each object a grade of membership ranging between zero and one. The notions of inclusion, union, intersection, complement, relation, convexity, etc., are extended to such sets, and various properties of these notions in the context of fuzzy sets are established. In particular, a separation theorem for convex fuzzy sets is proved without requiring that the fuzzy sets be disjoint.
>
> (Zadeh, 2015)

We are now turning to the critique of convex models of uncertainty. Bolotin (1968) writes:

> As an "antithesis" to the probabilistic approach, methods are occasionally put forward based on the concepts of "infrequently-occurring", "maximum" and "minimum" loads and strengths. In reality these are, however, no more than substitute probabilistic procedures dispensing with the theory of probability—and in spite of their apparent simplicity and obviousness, entailing remdiate logical contradictions. They are inapplicable in practice without arbitrary decision—which render them implausible and inadequate to a large degree.

Another objection is raised against convex models of uncertainty—known also as min-max. info-gap, guaranteed estimation, optimization and anti-optimization techniques—is their apparent conservativeness.

"History teaches the continuity of the development of science. We know that every age has its own problems, which the following age either solves or casts aside as profitless and replaces by new ones." This is a citation from David Hilbert's (1900) Seminar Lecture "Mathematical Problems."

Final quotation is of Kenneth Ewart Boulding (1910–1993)—English-born American economist, educator, and interdisciplinary philosopher, from his *Ballad of Philosophers* (see his book, 1980):

Philosophers are never slow
To talk of what they do not know.
They range the countries of the Mind
In search of what they cannot find,
Dashing along in hot pursuit
Of the elusive Absolute
Oblivious of the fact, one fears,
That it has been extinct for years.

Yes, there is a paradox! How can one discuss uncertainty, that which is not known with precision?!

Still, engineers, as pragmatic creatures, discern three competing and complementary approaches to uncertainty. Which one to choose depends on the amount and character of information available at hand. It is our firm belief that any of the described approaches is useful and valid.

Thus, an engineer is advised to choose the technique that engineer knows, without discarding or denigrating other techniques. It is best to got familiar with all three approaches, and utilize them, or their combination(s), in accordance with need and adopted philosophy within the organization.

Appendix Definitions for Terms Used in the Book

Action A set of concentrated or distributed forces acting on a structure (direct action), or deformations imposed on a structure or constrains within it (indirect action); the term "load" is also used to describe direct actions.

Basic variables Variables representing physical quantities that characterize actions, environmental effects, material properties, soil properties, geometry.

Capacity The ability to withstand external actions without failure, e.g. load bearing capacity, deformation capacity, rotation capacity.

Characteristic value A value chosen either on a statistical base, so that it can be considered to have prescribed probability of being exceeded toward unfavorable values; or on acquired experience or on physical constraints (i.e. nominal value).

Collapse Destruction of a part of structure.

Combination of actions (load combination) A set of design values used for verification of structural reliability for a limit state under different actions occurring simultaneously.

Design criteria Quantitative formulations describing the conditions to be complied with for each limit state.

Design service (working) life The period for which a structure or a structural element is assumed in design to be used for its intended purpose with expected maintenance but without major structural repair being necessary.

Design situation The set of conditions for which the design is required to demonstrate that relevant limit states are not exceeded during a specific tine period.

Design value The value of a basic variable used in a design criterion. Note: either multiplying or dividing a characteristic value by a factor obtains this value.

Deterministic method A calculation method in which all basic variables are treated as non-random.

Durability The ability of a structure or a structural element to maintain adequate performance for a given time under expected action and environmental influences.

Effects of actions All types of response to actions, such as internal forces and moments, stresses, deformations, cracking.

Environmental influences Chemical, biological, or physical influences (other than actions) which degrade the material of a structure thereby likely to affect its serviceability and safety in an unfavorable way.

Form of structure A particular arrangement of structural elements forming frames, suspended beams, arches, etc.

Geometrical imperfections Deviations from the intended geometry of a structure.

Hazard An exceptionally unusual and severe event, e.g. an abnormal action or environmental effect, insufficient strength or resistance, or excessive deviation from intended dimensions.

Importance factor A factor by accounting for the importance of failure implications for the given structure.

Limit states States beyond which the structure no longer satisfies the design performance requirements.

Limit states function A function of basic variables whose zero values are used to characterize a limit state.

Limit states method A calculation method intended to prevent a structure from exceeding specified limit states.

Maintenance A set of activities to be performed during the design service life of a structure in order to enable its reliable operation.

Mechanical failure Failure due to rupture, buckling, formation of a mechanism, excessive yielding.

Method of partial factors A calculation method accounting for the uncertainties and variability of basic variables by means of representative values, partial factors and, if relevant, additive quantities.

Model uncertainties Uncertainties related to accuracy of a model, including physical uncertainties, statistical uncertainties, etc.

Nominal value A value chosen on a non-statistical base, e.g. on acquired experience or on physical constraints.

Partial factors Reliability factors applicable to actions, internal forces and moments, material properties or resistances, and reflecting their uncertainties.

Permissible (allowable) stresses method A calculation method in which the stresses occurring under the expected maximum loads are compared with some fraction of the resistance of the material.

Probabilistic methods Calculation methods in which the relevant basic variables are treated as random (this term covers both reliability index methods and fully probabilistic methods).

Reference period A chosen period of time used as basis for assessing statistically variable actions, as well as for accidental actions.

Reliability The ability of a structure or a structural element to meet the specified requirements, including the service life, for which it is designed (it usually covers structural safety and serviceability, and can be expressed in terms of probability).

Reliability class The class of a structure or a structural element indicating the required level of reliability.

Reliability elements Numerical quantities used in the partial factors format, whereby the specified degree of reliability is supposedly reached.

Reliability index An indication of the probability of failure. A substitute for the failure probability pf defined as β = -P-1 (pf), where P-1 is the inverse standardized normal distribution.

Resistance The ability of a cross-section or a member of a structure to withstand actions without mechanical failure; e.g. bending resistance, buckling resistance, tension resistance.

Risk The danger that an undesired event represents for humans, environment, or properties; risk is expressed in terms of possible consequences of the undesired event (cost of losses, for example) and associated probabilities.

Robustness or structural insensitivity The ability of a structure to withstand events like fire, explosion, impact, or consequences of human errors, without being damaged to an extent disproportionate to the original state.

Rupture Total loss of strength of a structural element or a part of it.

Serviceability The ability of a structure or a structural element to perform adequately in normal service under expected actions.

Serviceability limit states States beyond which the specified service requirements are no longer met.

Statistical uncertainties Uncertainties related to the values of statistical parameters, or to choose of the statistical distributions of basic variables.

Stiffness A property relating deformation to the action causing it.

Strength A mechanical property of material expressed usually in units of stress.

Structural element A physically distinguishable part of structure, e.g. column, beam, slab, shell, etc.

Structural failure Exceeding any specified performance criterion.

Structural response The behavior of a structure, or a structural element or part of it, or material under certain actions and/or certain environmental influences.

Structural safety The ability of a structure or a structural element to resist the expected action (including certain accidental phenomena) that it has to withstand during construction and anticipated service (ultimate limit state related).

Structural system Structural elements of a construction work and the way these elements function together.

Structure An organized combination of connected parts designed to provide some measure of resistance and rigidity against various actions.

Survivability Preservation of the ability of a structural system to carry out its main functions under accidental actions without progressive collapse or cascade development of failures.

Ultimate limit states States associated with collapse, or with other similar forms of structural failure (they generally correspond to the maximum load-bearing resistance).

Bibliography

There is nothing more instructive than a good bibliography.

G. Sarton, 1952

Abbas M. (ed.) (2012) *Earthquake-Resistant Structures: Design, Assessment and Rehabilitation*, Croatia: InTech.

Abdel-Tawab K. and Noor A.K. (1999) A Fuzzy Set Analysis for a Dynamic Thermo-Elasto-Viscoplastic Damage Response, *Computers & Structures*, Vol. 70 (1), 91–107.

AIJ (1996) *Recommendations for Loads on Buildings*, Tokyo: Architectural Institute of Japan.

Aitkin C.G.G. and Taroni F. (2004) *Statistics and the Evaluation of Evidence for Forensic Scientists (Statistics in Practice)*, New York: Wiley.

Ang A.H-S. and Amin M. (1969) Safety factors and Probability in Structural Design, *Journal of Structural Division*, Vol. 95, 1389–1404.

Ang A.H-S. and De Leon D. (1997) Determination of Optimal target Reliabilities for Design and Upgrading Structures, *Structural Safety*, Vol. 19, 91–103.

Apostolakis G. and Kaplan S. (1965) Pitfalls in Risk Calculations, *Reliability Engineering*, Vol. 2, 135–145.

Arbocz J. (1982) The Imperfection Data bank: A Means to Obtain Realistic Buckling Loads, in *Buckling of Shells* (E. Ramm, ed.), Berlin: Springer, 535–567.

Arbocz J. and Abramovich H. (1979) The Initial Imperfection Data Bank at the Delft University of Technology, Report LR-290.

Arnold V.I. (1997) *Selected Works*, Moscow: FASIC Publ. House (in Russian).

Arora J.S. (1989) *Introduction to Optimum Design*, New York: McGraw-Hill.

Arutjunjan N. Ch. (1952) *Some Questions of the Theory of Creep*, Moscow: "GITTL" Publishers (in Russian).

Augusti G., Baratta A., and Casciati F. (1979) Structural Response under Random Uncertainties, *Proceedings of the 3rd International Conference on Applications of Statistics and Probability in Soil and Structural Engineering (ICASP3)*, Sydney, Vol. 3, 280–328.

Augusti G., Baratta A., and Casciati F. (1984) *Probabilistic Methods in Structural Engineering*, London: Chapman & Hall.

Ayyub B.M.(ed.) (1997) *Uncertainty Modelling and Analysis in Civil Engineering*, Boca Raton: CRC Press.

Barlow R.E. and Proschan F. (1975) *Statistical Theory of Reliability and Life Testing Probability Models*, New York: Holt, Rinehart & Winston.

Bedford T. and Cooke R. (2001) *Probabilistic Risk Analysis: Foundations and Methods*, Cambridge, UK: Cambridge University Press.

BeicheltF. andFrankenP. (1983) *ZuverlässigkeitundInstadhaltung: MathematischeMethoden*, Berlin: VEB Verlag Technik (in German).

De Bélidor Bernard Forest (1729) *La science des ingénieurs, dans la onduit des travaux de fortification et d'architecture civile*, Paris: Chez Claude Jombert (in French).

Benaroya H., Nagurka M., and Han S. (2017) *Mechanical Vibration: Analysis, Uncertainties, and Control*, Boca Raton: CRC Press.

Bendat J.S. and Piersol A.G. (1971) *Random Data: Analysis and Measurement Procedures*, New York: Wiley.

Ben-Haim Y. (1985) *The Essay of Spatially Random Material*, Dordrecht: D. Reidel Publishing Company.

Ben-Haim Y. (1994) Convex Models of Uncertainty: Applications and Implications, *Erkenntnis: An International Journal of Analytic Philosophy*, Vol. 41, 139–156.

Ben-Haim Y. (1996) *Robust Reliability in the Mechanical Sciences*, Berlin: Springer.

Ben-Haim Y. (1997) Robust Reliability of Structures, in *Advances in applied Mechanics* (J.W. Hutchinson and T.Y. Wu, eds.), San Diego, CA: Academic Press, Vol. 33, 1–41.

Ben-Haim Y. (2001) *Information-Gap Decision Theory*, San Diego, CA: Academic Press.

Ben-Haim Y. (2010) *Info-Gap Economics: An Operational Introduction*, New York: Palgrave Macmillan.

Ben-Haim Y. and Elishakoff I. (1989) Dynamics and Failure of a Thin Bar with Unknown but Bounded Imperfections, in *Recent Advances in Impact Dynamics of Engineering Structures* (D. Hui and N. Jones, eds.), AMD Vol. 105, 89–96.

Ben-Haim Y. and Elishakoff I. (1990) *Convex Models of Uncertainty in Applied Mechanics*, Amsterdam: Elsevier Science Publishers.

Benjamin J.R. and Cornell C.A. (1970) *Probability, Statistics and Decision for Civil Engineers*, New York: McGraw Hill.

Ben-Tal A., El Ghaoui L., and Nemirovski A. (2009) *Robust Optimization*, Princeton, NJ: Princeton University Press.

Ben-Tal A. and Nemirovski A. (1998) Robust Convex Otimization, *Mathematics of Operations Research*, Vol. 23 (4), 769–805.

Berg A.I. (1961) Science about Reliability, *Economic Gazette*, 8 June (in Russian).

Bernstein S.A. (1957) *Essays on History of Structural Mechanics*, Moscow: Sate Publishing House on Civil Engineering and Architecture (in Russian).

Bertsimas D., Brown D.B. and Caramanis C. (2011) Theory and Applications of Robust Optimization, *SIAM Review*, Vol. 53, Issue (3), 464–501.

Birger I.A. (1970) Probability of Failure, Safety Factors and Diagnosis, in *Problems of Mechanics of Solid Bodies*, Leningrad: Sudostroenie Publishing House, 71–82 (in Russian).

Blekhman I.I., Myshkis A.D., and Panovko Ya. G. (1990) *Mechanics and Applied Mathematics: The Logics and Specificities of Mathematics' Applications* (second edition), Moscow: "Nauka" Publishing House (in Russian).

Blockley D.I. (1992) *Engineering Safety*, New York: McGraw-Hill.

Bodner S. (2003) Private Communication to I.E., September.

Bolotin V.V. (1967) Statistical Aspects in the Theory of Structural Stability, in *Dynamic Stability of Structures* (G. Herrmann, ed.), Oxford: Pergamon Press, 67–81.

Bolotin V.V. (1969) *Statistical Methods in Structural Mechanics*, San Francisco: Holden Day.

Bolotin V.V. (1968) Modern Status of the Reliability Theory and Statistical Mechasnics of Structures, in *Reliability Problems in Structural Mechanics,* (V.V. Bolotin and A. Chyras, eds), Vilnius: "Vaidaz" Publishing house, 7–13 (in Russian).

Bolotin V.V. (1984) *Random Vibrations of Elastic Systems*, The Hague: Martinus Nijhoff Publishers.

Boulding K.E. (1980) *Beasts, Ballads, and Bouldingisms*, New Brunswick: Tarnsaction Books, 76.

Bruhn E.F. (ed.) (1967) *Analysis and Design of Missile Structures*, Cincinnati, OH: Tri-State offset Company.

Bruhn E.F. (1993) *Analysis and Design of Flight Vehicle Structures*, New York: McGraw-Hill.

Bushnell D. (1990) GENOPT – A Program That Writes User Friendly Optimization Code, *International Journal of Solids and Structures*, Vol. 26 (9–10), 1173–1210.

Cai K., Wen C., and Zhang M. (1991) Fuzzy Variables as a Basis for the Theory of Fuzzy Reliability in the Possibility Context, *Fuzzy Sets and Systems*, Vol. 42, 145–172.

Calafiore G. and Dabbene F. (eds.) (2006) *Probabilistic and Randomized Methods for Design under Uncertainty*, Berlin: Springer.

Carter A.D.S. (1997) *Mechanical Reliability and Design*, New York: Wiley.

Casciati F. (1991) Safety Index, Stochastic Finite Elements and Expert Systems, in *Reliability Problems: General Principles and Applications in Mechanics of Solids and Structures* (F. Casciati and J.B. Roberts, eds.), Vienna: Springer, 51–88.

Citaku E. (2020) *What If? Your Guide to Making the Best Decisions Ever*, Ohio: Citaku Publishing.

Cherkesov G.N. (1987) *Methods and Models of Survivability Assessment for Complex Systems*, Moscow: "Znanie" Publishing House (in Russian).

Coleman S.Y. (2013) Statistical Thinking in the Quality Movement ±25 Years, *The TQM Journal*,

Cooke R., Bedford T., Meilijson I., and Meester L. (1993) Design of Reliability Data bases for Aerospace applications, European Space Agency, Delft UT, Technical Report, 93–110.

Cooke R. (1996) The Design of Reliability Data Bases, Part II: Competing Risk and Data Compression, *Reliability Engineering & System Safety*, Vol. 51 (2), 209–223.

Cornell C.A. (1969) Bounds on the Reliability of Structural Systems, *Journal of Structural Division*, Vol. 93, STI, 171–200.

Cramer D. (2003) *Advanced Quantities Data Analysis*, London: McGraw-Hill Education (UK).

Cramer H. and Leadbetter M.R. (2004) *Stationary and Related Stochastic Processes*, Mineola, NY: Dover Publications.

Cursi E.S. and Sampao R. (2015) *Uncertainty Quantification and Stochastic Modeling with Matlab*, London: ISTE-Wiley.

DeFinetti B. (1974) *Theory of Probability: A Critical Introductory Treatment*, Vol. 1, New York: Wiley.

Dempster A.P. (1967) Upper and Lower Probability Induced Multivalued Mapping, *The Annals of Mathematical Statistics*, Vol. 38 (2), 325–339.

Ditlevsen O. (1979) Narrow Reliability Analysis of Frame Structure, *Journal of Structural Mechanics*, Vol. 1 (4), 453–472.

Ditlevsen O. (1981) *Uncertainty Modeling with Applications to Multidimensional Systems*. New York: McGraw-Hill.

Ditlevsen O. (1988) *Uncertainty and Structural Reliability: Hocus Pocus or Objective Modelling?* Afdelingen for BærendeKonstruktioner, DanmarksTekniskeHøjskole.

Ditlevsen O. and Arnbjerg-Nielsen T. (1989) Decision Rules in Re-evaluation of Existing Structures, *Proceedings, DABI Symposium on Re-evaluation of Concrete Structures*, Danish Concrete Institute, 239–248.

Ditlevsen O. and Madsen H.O. (1996) *Structural Reliability Methods*, New York: Wiley.

Doorn N. (2021) Sven Ove Hansson's Contribution to Philosophy of Technology and Engineering, available at https://philarchive.org/archive/DOOSOH, accessed 26 April.

Doorn N. and Hansson S. O. (2011). Should Probabilistic Design Replace Safety Factors? *Philosophy & Technology*, Vol. 24, (2), 151–168.

Doorn N. and Hansson, S.O. (2015) Design for the Value of Safety, in *Handbook of Ethics and Values in Technological Design* (van den Hoven J., Vermaas P.E. and van de Poel I, eds.), New York: Springer, 491–509.

Doorn N. and Hansson S.O. (2018) Factors and Margins of Safety, *Handbook of Safety Principles* (Möller N., Hansson S.O. and Holmberg J-E., eds.), New York: Wiley, 87–114.

Dresden M. (1992) Chaos: A New Scientific Paradigm or Science by Public Relations? *The Physics Teacher*, Vol. 30, 10–14 and 74–80.

Driving A.Y. (1972) Recommendation for Applying Economical Methods in Design of Structures with Economical Responsibility (in Russian). Proceedings of TSNIISK, Moscow (in Russian).

Druzhinin T.V. (1977) *Reliability of Automatic Systems*, Moscov: "Energia" Publishing House (in Russian).

Dubois D. and Prade H. (1985) Evidence Measures Based on Fuzzy Information, *Automatica*, Vol. 21 (5), 547–562.

El Hami A. and Radi B. (2013) *Uncertainty and Optimization in Structural Mechanics*, London: ISTE-Wiley.

Elishakoff I. (1995) Essay on Uncertainties in Elastic and Viscoelastic Structures: From AM Freudenthal's Criticisms to Modern Convex Modeling, *Computers & Structures*, Vol. 56 (6), 871–895.

Elishakoff I. (ed.) (1999) *Whys and Hows in Uncertainty Modelling: Probability, Fuzziness and Anti-optimization*, Vienna: Springer.

Elishakoff I. (1999) *Probabilistic Theory of Structures*, New York: Dover (first edition: Wiley, 1983).

Elishakoff I. (2003) Notes on the Philosophy of the Monte Carlo Method, *International Applied Mechanics*, Vol. 39 (7), 3–14.

Elishakoff I. (2004) *Safety Factors and Reliability: Friends or Foes?* Dordrecht: Kluwer Academic Publishers.

Elishakoff I. (2004) *Safety Factors and Reliability: Friends or Foes?* Dordrecht: Kluwer Academic Publishers.

Elishakoff I. (2017) *Probabilistic Methods in the Theory of Structures*, Singapore: World Scientific.

Elishakoff I. and Colombi P. (1993) Combination of Probabilistic and Convex Models of Uncertainty when Scarce Knowledge is Present on Acoustic Excitation Parameters, *Computer Methods in Applied Mechanics and Engineering*, Vol. 104, 187–209.

Elishakoff I. and Ferracuti B. (2006a) Fuzzy Sets-Based Interpretation of the Safety Factor, *Fuzzy Sets and Systems*, Vol. 157 (18), 2495–2512.

Elishakoff I. and Ferracuti B. (2006b) Four Alternative Definitions of the Fuzzy Safety Factor, *Journal of Aerospace Engineering*, Vol. 19 (4), 281–287.

Elishakoff I., Gana-Shvili Y., and Givoli D. (1991) Treatment of Uncertain Imperfections as a Convex Optimization Problem, Proceedings of the Sith International Conference on Applications of Statistics and Probability in Civil Engineering (L. Esteva and S.E. Ruiz, eds.), Mexico, Vol. 1, 150–157.

Elishakoff I. and Hasofer A.M. (1985) On the Accuracy of Hasofer-Lind Reliability Index, in *Proceedings of ICOSSAR-85, International Conference on Structural Safety and Reliability*, Kobe, Japan, 1/229–1/239.

Elishakoff I. and Hasofer A.M. (1987) Exact Versus Approximate Analyses in Structural Reliability, *Applied Scientific Research, International Journal of Thermal, Mechanical and Electromagnetic Phenomena in Continua*, Vol. 44, 303–312.

Elishakoff I. and Li Q. (1999) How to Combine Probabilistic and Antioptimization methods? in *Whys and Hows in Uncertainty Modeling*, Vienna: Springer, 319–340.

Elishakoff I., Li Y-W., and Starnes J.H. Jr. (2001) *Non-Classical Problems in the Theory of Elastic Stability*, Cambridge, UK: Cambridge University Press.

Elishakoff I., Lin Y-K., and Zhu L-P. (1994) *Probabilistic and Convex Modeling of Acoustically Excited Structures*, Amsterdam: Elsevier Science Publishers.

Elishakoff I. and Nordstrand T. (1991) Probabilistic Analysis of Uncertain Eccentricities in a Model Structure, Proceedings of the Sixth Int'l Conference on Applications of Statistics and Probability in Civil Engineering (L. Esteva and S.E. Ruiz, eds.), Mexico City, Vol. 1, 184–192.

Elishakoff I. and Ohsaki M. (2010) *Optimization and Anti-optimization of Structures under Uncertainty*, London: Imperial College Press.

Elochin A.N. (1994) To the Problem of Determining Acceptable Risk Criteria, in *Safety Problems at Accidental Situations*, Issue 8, 42–50, Moscow (in Russian).

Faber M.H. (2005) On the Treatment of Uncertainties and Probabilities in Engineering Decision Analysis, *Journal of Offshore Mechanics and Arctic Engineering*, Vol. 127, 243–248.

Faber M.H. (2008) *Risk Assessment in Engineering-Principles, System Representation and Risk Criteria*, Zurich: ETH Zurich.

Feld J. (1964) *Construction Failure*, New York: Wiley.

Feller W. (1970) *An Introduction to Probability Theory and its Applications*, New York: Wiley.

Ferry-Borges J. and Castanheta M. (1971) *Structural Safety*, Lisbon: LNEC.

Feynman R. P (2005) *The Pleasure of Finding Things Out: The Best Short Works of Richard P. Feynman*, New York: Basic Books.

Fischer D.H. (1970) *Historian's Fallacies*, New York: Harper & Row, 3–8.

Flint A.R. (1981) Risks and Their Control in Civil Engineering, *Proceedings of the Royal Society of London: A. Mathematical and Physical Sciences*, Vol. 376 (1764), 167–179.

Florin V.A. (1951) *Fundamentals of Soil Mechanics*, Vol. 2, Moscow: State Publishing in Civil Engineering (in Russian).

Freudenthal A.M. (1946) The Statistical Aspect of Fatigue of Materials, *Proceedings of the Royal Society of London. Series A. Mathematical and Physical Sciences*, Vol. 187 (1011), 416–429.

Freudenthal A.M. (1947) Safety of Structures, *TransactionsASCE,* Vol. 112, 125–180.

Freudenthal A.M. (1948) Reflections on Standard Specifications for Structural Design, *Transactions of the American Society of Civil Engineers*, Vol. 113 (1), 269–287.

Freudenthal A.M. (1956) Safety and the Probability of Structural Failure, *Transactions of the American Society of Civil Engineers*, Vol. 121 (1), 1337–1375.

Freudenthal A.M.(1957) *Safety and Safety Factor for Airframes, Report 153, NATO*, Paris: Palais de Chaillot.

Freudenthal A.M. (1961) Safety, Reliability and Structural Design, *Journal of the Structural Division*, Vol. 87 (3), 1–16.

Freudenthal A.M. (1972) Reliability Analysis Based on Time to the First Failure, in *Aircraft Fatigue* (Symposia/International Committee on Aeronautical Fatigue), Oxford: Pergamon Press, 13–48.

Freudenthal A.M. (1974) New Aspects of Fatigue and Fracture Mechanics, *Engineering Fracture Mechanics*, Vol. 6 4), 775–780.

Freudenthal A.M. (1977) The Scatter Factor in the Reliability Assessment of Aircraft Structures, *Journal of Aircraft*, Vol. 14 (2), 202–208.

Freudenthal A.M (1981) *Selected Papersby Alfred M. Freudenthal*, Civil Engineering Classics, New York: ASCE Press.

Freudenthal A.M., Garretts J.M. and Shinozuka M. (1966) The Analysis of Structural Safety, *Journal of the Structural Division*, Vol. 92 (1), 267–326.

Freudenthal A. M. and Gumbel E. J. (1953) On the Statistical Interpretation of Fatigue Tests, *Proceedings of the Royal Society of London. Series A. Mathematical and Physical Sciences*, Vol. 216 (1126), 309–332.

Freudenthal A.M. and Gumbel E.J. (1954) Failure and Survival in Fatigue, *Journal of Applied Physics*, Vol. 25 (11), 1435–1435.

Freudenthal A.M. and Gumbel E.J. (1956) Physical and Statistical Aspects of Fatigue, in *Advances in Applied Mechanics*, Vol. 4 Amsterdam: Elsevier, 117–158.

Freudenthal A.M. and Schuëller G.I. (1976) Risk Evaluation for Structures, *Konstruktiver Ingenieurbau Berichte*, 7–95.

Galambos J. (1987) *The Asymptotic Theory of Extreme Order Statistics*, Malabar, FL: Robert Krieger Publishers.

Galilei G. (1638) *Dialogues Concerning Two New Sciences*, Buffalo, NY: Prometheus Books (1991 edition).

Geniev G.A. (1993) On Assessing Dynamic Effects in Brittle Bar Systems, *Journal of Concrete and Reinforced Concrete*, (3), 71–72, Moscow (in Russian).

Geniev G.A. (2000) Application of Direct Mathematical Methods to Optimization of Reliability Characteristics of Combined Systems, *Proceedings, University of Civil Engineering*, Vol. 1, Moscow, 16–21 (in Russian).

Gertsbakh I.B. and Kordonsky Kh. B. (1969) *Models of Failure*, Berlin: Springer.

Gil-Aluja J. (2001) *Handbook of Management under Uncertainty*, Berlin: Springer.

Gnedenko B.V., Belyaev Y.K., and Soloviev A.D. (1969) *Mathematical Methods in the Theory of Reliability*, New York: Academic Press.

Gnedenko B.V., Pavlov I.V., and Ushakov I.A. (S. Chakravarthy, ed.) (1999) *Statistical Reliability Engineering*, New York: Wiley.

Gnedenko B.V. and Ushakov I.A. (1995) *Probabilistic Reliability Engineering*, New York: Wiley.

Goldstein H. (2002) *Classical Mechanics*, London: Addison-Wesley Publishing.

Good I.J. (1995) Reliability Always Depends on Probability of Course, *Journal of Statistical Computation and Simulation*, Vol. 52, 192–193.

Good I.J. (1996) Reply to Prof. Y. Ben-Haim, *Journal of Statistical Computation and Simulation*, Vol. 55, 265–266.

Gore A. (1995) The National Information Infrastructure, *Journal of Computers and Science Teaching*, Vol. 14 (172), 27–33.

Gross J.L. and Yellen J. (2005) *Graph Theory and its Applications*, Boca Raton: CRC Press.

Gumbel E.L. (1960) Bivariate Exponential Distributions, *American Statistical Association Journal*, Vol. 55 (292), 698–707.

Gumbel E.J. (1967) *Statistics of Extremes*, New York: Columbia University Press.

Hamming R.W. (1987) *Numerical Methods for Scientists and Engineers*, New York: Dover Publications.

Hansson S.O. (2007) Philosophical Problems in Cost–Benefit Analysis, *Economics & Philosophy*, Vol. 23 (2), 163–183.

Hansson S.O. (2009a) Risk and Safety in Technology, in *Philosophy of Technology and Engineering Sciences*, North-Holland, 1069–1102.

Harris B. and Soms A.P. (1983) A Note on a Difficulty Inherent in Estimating Reliability from Stress–Strength Relationship, Naval Research Logistics Quarterly, Vol. 30 (4), 659–663.

Hart G.C. (1982) *Uncertainty Analysis, Loads, and Safety in Structural Engineering*, Englewood Cliffs, NJ: Prentice Hall.

Hart G.C. (1982) Uncertainty Analysis: Loads, and Safety in structural Engineering, Englewood Cliffs, NJ: Prentice Hall.

Hasofer A.M. and Lind N.C. (1974) An Exact and Invariant First Order Reliability Format, *Journal of Engineering Mechanics*, Vol. 100, 111–121.

Haugen E.B. (1968) *Probabilistic Approaches to Design*, New York: Wiley.

Haugen E.B. (1980) *Probabilistic Mechanical Design*, New York: Wiley-Ineterscience.

Heisenberg W. (1958) *Physics and Philosophy: The Revolution in Modern Science*, New York: Harper, 58.

Hilbert D. (1900) Mathematische Probleme, *Nachrichten von der Koniglichen Gesellschaft der Wissenschaftenzu Gottingen* (in German).

Hlaváček I., Chleboun J., and Babuška I. (2004) *Uncertain Input Data Problems and the Worst Scenario Method*, Amsterdam: Elsevier.

Huber P.J. (1964) Robust Estimation of a Location Parameter, in *Breakthroughs in Statistics*, New York: Springer, 492–518.

JCSS (2005) Joint Committee of Structural Safety, Probabilistic Model Code, available at publication, www.jscc.ethr.ch.

Jennings, B. (2011) available at www.quantumdiaries.org/2011/10/28/scientists-groping-in-the-dark/

Jennings B.K. (2015) *In Defense of Scientism: Insider's View of Science*, MUSQUOD.

Jensen F. (2000) It Mustn't Break (editorial), *Reliability Engineering International,* Vol. 16, 1.

Johnson A.I. (1953) Strength, Safety and Economical Dimensions of Structures, *Bulletin, Division of Structural Engineering,* Royal Institute of Technology, Stockholm, No. 12.

Kálmán R. E. (1972), cited in: Lotfi A. Zadeh (2004) Fuzzy Logic Systems, Origin, Concepts and Trends http://wi-consortium.org/wicweb/pdf/Zadeh.pdf November 10, 2004 (accessed on July 11, 2021). (Source: https://quotepark.com/authors/rudolf-e-kalman/).

Kanda J. (1996) Normalized Failure Cost as a Measure of a Structural Importance, *Nuclear Engineering and Design*, Vol. 160, 299–305.

Kapur K.S. and Lamberson L.R. (1977) *Reliability in Engineering Design*, New York: Wiley.

Kaufmann A. and Gupta M.M. (1985) *Introduction to Fuzzy Arithmetic: Theory and Applications*, New York: Van Nostrand.

Kettelle J.D. (1962) Least-Cost Allocation of reliability Investment, *Operations Research*, Vol. 10, 249–265.

Khozialov N.F. (1929) Safety Factor in Strength, *Journal of Construction Industry*, (10), 840–844 (in Russian).

Klir G.J. (1994) On the Alleged Superiority of Probabilistic Representation of Uncertainty, *IEEE Transactions on Fuzzy Systems*, Vol. 2 (1), 27–31.

Klir G.K., St. Clair U., and Yuan B. (1997) *Fuzzy Set Theory: Foundations and Applications*, New York: Prentice Hall PTR.

Klir G.J. and Yuan B. (1995) *Fuzzy Sets and Fuzzy Logic: Theory and Applications*, Englewood Cliffs, NJ: Prentice Hall.

Knoll F. (1976) Commentary on the Basic Philosophy and Recent Development of Safety Margins, *Canadian Journal of Civil Engineering*, Vol. 3 (3), 409–416.

Körner R. (1997) On the Variance of Fuzzy Random Variables, *Fuzzy Sets and Systems*, Vol. 92, 83–93.

Kosko B. (1993) *Fuzzy Thinking*, New York: Hyperion, Chapter 1, 12.

Kosko B. (1994) The Probability Monopoly, *IEEE Transactions on Fuzzy Systems*, Vol. 2 (1), 32–33.

Krick E.V. (1969) *Engineering & Engineering Design* (second edition), New York: Wiley.

Kudsis A.P. (1985) *Reliability Estimation of Reinforced Concrete Structures*, Vilnius: Moklas Publishing House.

Laviolette J.M. and Seaman J.W., Jr. (1994) The Efficacy of Fuzzy Representations of Uncertainty, *IEEE Transactions on Fuzzy Systems*, Vol. 2 (1), 4–15.

Lemaire M. (2009) *Structural Reliability*, London: ISTE-Wiley.

Leunberger D.G. (1984) *Introduction to Linear and Non-Linear Programming,* Reading, MA: Addison-Wesley.

Lew H.S. (2005) *Best Practice Guidelines for Mitigation of Building Progressive Collapse*. Building and Fire Research Laboratory, NIST, Gaithersburg, MD, 20899–8611.

Lodwick W.A. and Thipwiwatpotjana P. (2017) *Flexible and Generalized Uncertainty Optimization Theory and Methods*, Cham: Springer.

Lorkowski J. and Kreinovich V. (2017) *Bounded Rationality in Decision Making under Uncertainty: Towards Optimal Granularity*, Cham: Springer.

Lovelace A.M. (1972) Keynote Address, in *International Conference on Structural Safety and Reliability* (A.M. Freudenthal, ed.), Oxford: Pergamon Press, 4.

Lyons D. (2013) *Optimization under Uncertainty with Applications to Multi-Agent Coordination*, Germany: KIT-Bibliothek.

Madsen H.O. (1987) Model Updating in Reliability Theory, in *Reliability and Risk Analysis in Civil Engineering, Proceedings, ICASP 5* (N.C. Lind, ed.), Institute for Risk Research, University of Waterloo, Vol. 1, 564–577.

Madsen H.O., Krenk S., and Lind N.C. (2006) *Methods of Structural Safety*, Mineola: Dover Publications.

Mahmoudi, M. (2020) A Missing, but Essential, Platform for Multidisciplinary Scientific Discussion: Understanding the "Elephant," *Future Science OA*, 7 (3), FSO66.

Mattila J.K. (2004) On Aristotelian Tradition and Fuzzy Revolution, in *2004 IEEE International Conference on Fuzzy Systems (IEEE Cat. No. 04CH37542)*, Vol. 3, 1585–1588, New York: IEEE Press.

Mayer M. (1926) *Die Sicherheit der Bauwerke und ihre Berechnung nach Granzkrafte nstatt nach Zulassigen Spannungen*, Berlin: Springer (in German).

Maymon G. (1998) *Some Engineering Applications in Random Vibrations and Random Structures,* Reston, VA: AIAA Press.

Maymon G. (2000) The Stochastic Safety Factor – A Bridge Between Deterministic and Probabilistic Structural Analysis, *Frontiers Science Series*, 693–700.

Maymon G. (2002b) Personal Communication to I.E., 26 September.

Maymon G. (2008) *Structural Dynamics and Probabilistic Analysis for Engineers*, London: Butterworth-Heinemann.

Melchers R.E. and Beck A.T. (2018) *Structural Reliability Analysis and Prediction*, New York: Wiley.

Metcalfe A. H. (1961) Discussion of "Philosophical Aspects of Structural Design", *Journal of the Structural Division*, Vol. 87 (8), 249.

Middleton D. (1996) *An Introduction to Statistical Communication Theory*, New York: Wiley-IEEE Press (Classic Reissue).

Mkrtychev O.V. and Raiser V.D. (2016) *Theory of Reliability in Structural Design*, Moscow: ASU Publishers (in Russian).

Moses F. (1997) Problems and Prospects of Reliability-Based Optimization, *Engineering Structures*, Vol. 19 (4), 293–301.

Möller B. and Beer M. (2004) *Fuzzy Randomness: Uncertainty in Civil Engineering and Computational Mechanics*, Berlin: Springer.

Murzewski J. (1963) *The Introduction to the Structural Safety Theory*, Warsaw: Panstwowe WydawnictwoNaukowe (in Polish).

Murzewski J. (1970) *BezpieczenstwoKonstrukejiBudowlanych*, Warsaw: "Arkady" Publishing House (in Polish).

Natke H.G. and Ben-Haim Y. (1997) *Uncertainty: Models and Measures*, Wiley Australia.

Navier C.L.M.H. (1827) *Mémoire sur les lois de l'équilibre et du mouvement des corps solides élastiques* (in French).

Newson R.J. (1962) Discussion of "Philosophical Aspects of Structural Design", *Journal of the Structural Division*, Vol. 88 (1), 197–199.

NKB (1978) Recommendations for Loading and Safety Regulations for Structural Design. Copenhagen Report No. 36, 64.

Noor A.K., Starnes J.H. Jr., and Peters J.M. (2000) Uncertainty Analysis of Composite Structures, *Computer Methods in Applied Mechanics and Engineering*, Vol. 185 (2–4), 413–432.

Noor A.K., Starnes J.H. Jr., and Peters J.M. (2001) Uncertainty Analysis of Stiffened Composite Panels, *Composite Structures*, Vol. 51 (2), 139–158.

Norton R.L. (2000) *Machine Design*, Upper Saddle River, NJ: Prentice Hall, 19–22.

Nowak A.S. and Collins K.R. (2013) *Reliability of Structures*, Boca Raton: CRC Press.

Otstavnov V.A., Smirnov A.F., Raizer V.D., and Sukhov Y.D. (1981) Introduction of Importance Factor in Building Code, *Journal of Structural Mechanics & Analysis of Structures*, (1), 11–14 (in Russian).

Ovchinnikov I.G. (1982) *Mathematical Corrosion Prognosis of Structural Steel Elements,* Moscow: VINITI typescript #2061 (in Russian).

Owhadi H. and Scovel C. (2017) Towards Machine Wald, in *Handbook of Uncertainty Quantification* (R. Ghanem, D. Higdon and H. Owhadi H., eds.), Cham: Springer, 151–191.

Parzen E. (1960) *Modern Probability Theory and its Applications*, New York: Wiley.

Paté-Cornell M.E. (1996) Uncertainties in Risk Analysis: Six Levels of Treatment, *Reliability Engineering & System Safety*, Vol. 54 (2–3), 95–111.

Pearson K., Quote available at https://quotefancy.com/quote/1598428/Karl-Pearson-The-record-of-a-month-s-roulette-playing-at-Monte-Carlo-can-afford-us (accessed on July 21, 2021).

Perelmuter A.V. (1995) Assessing the Survivability of Supporting Structures, in "Steel Structures", in *Works of Prof. Streletsky School* (A. Kudishin, ed.), Moscow: MGSU Publishing House, 62–68 (in Russian).

Perelmuter A.V. (1999) *Selected Problems of Reliability and Structural Safety*, Kiev: Ukrainian Institute of Steel Constructions Publishing House (in Russian).

Pereverzev E.S. (1987) *Stationary Processes in Parametric Models of Reliability*, Kiev: "Naukova Dumka" Publishing House.

Petroski H. (1996) *To Engineer Is Human: The Role of Failure in Successful Design*, New York: Vintage.

Pikovaky A.A. (1961) *Statics of Column Systems with Compressed Elements*, Moscow (in Russian).

Plot M. (1936) Noir Sur la Notion de Coefficient de Securite, *Annales des Ponts et Chaussées,*Paris, T II, fase 7 (in French).

Prigogine I. (2017) *Nonequilibrum Statistical Mechanics*, New York: Dover.

Pugsley A. (1966) *The Safety of Structures*, London: Arnold.

Qu X. and Haftka R.T. (2002) Response Surface Approach Using Probabilistic Safety Factor for Reliability – Based Design Optimization, *Proceedings, 2nd Annual probabilistic Method Conference*, Newport Beach, CA.

Rackwitz R. and Fiessler B. (1978) Structural Reliability under Combined Load Sequences, *Computers and Structures*, Vol. 9, 489–494.

Raizer V.D. (1985) Stochastic Instability of Parametric Oscillations of Nonlinear Systems, *Journal of Structural Mechanics and Analysis of Structures*, (4), 43–47 (in Russian).

Raizer V.D. (1986) *Reliability Theory Methods for Design Code Making*, Moscow: "Stroyizdat" Publishing House (in Russian).

Raizer V. (1990) Reliability of Structures Subjected to Corrosive Wear, *Journal of Technical Mechanics*, (11), Berlin.

Raizer V.D. (1995) *Analysis and Codification of Structural Reliability*, Moscow: "Stroyisdat" Publishing House (in Russian).

Raizer V.D. (2000) Probabilistic Analysis of Optimal Lengths between Thermal Joints, *Proceedings of the 8th ACSE Specialty Conference*, Univ Notre-Dame, IN.

Raizer V.D. (2001) Reliability Optimization for Pipelines to Corrosion Wear, *Proceedings of ASCE Conference, Advances in Pipeline Engineering & Construction, Pipelines* 2001, San Diego, CA.

Raizer V.D. (2004) Theory of Reliability in Structural Design, *Applied Mechanics Reviews*, Vol. 57 (1), 1–21.

Raizer V.D. (2009) *Reliability of Structures: Analysis and Applications*, North Vancouver: Backbone Publishers.

Raizer V.D. (2010) *Theory Reliability of Structures*, Moscow: ASU Publishing House (in Russian).

Raizer V.D. (2018) *Probabilistic Methods in Analysis of Structural Reliability and Survivability*, Moscow: ASU Publishing House (in Russian).

Raizer V.D. and Mkrtychev O.V. (2000) Nonlinear Probabilistic Analysis for Multiple-unit Systems, *Proceedings, 8th ASCE Specialty Conference on Probabilistic Mechanics*, University of Notre Dame.

Raizer V.D., Sukhov Y.D., and Lebedeva I.V. (1998) Reliability Analysis of Multiplex Spatial Structures, *Proceedings, IASS Congress.*

Randall F.A. (1977) Historical Notes on Structural Safety, *Proceedings, American Concrete Institute*, Vol. 70, 669–679.

Rao S.S. (1992) *Reliability-Based Design*, New York: McGraw-Hill.

Roark R.J. (1965) *Formulas for Stress and Design*, New York: McGraw-Hill.

Roësset J. and Yao J.T.P. (1988) Civil Engineering Needs in the 21st Century, *Journal of Professional Issues Engineering*, Vol. 114 (3), 248–255.

Roësset J. and Yao J.T.P. (2002) State of the Art of Structural Engineering, *Journal of Structural Engineering*, Vol. 128 (8), 965–975.

Rosenblueth E. (1975) Design Philosophy: Structures, in *Proc. 2nd International Conference on Application of Statistics and Problems in Soil and Structural Engineering*, Aachen, West Germany.

Rosenblueth E. (1987) What Should We Do with Structural Reliabilities? in *Reliability and Risk Analysis* (N.C. Lind, ed.), Ontario: University of Waterloo Press, 24–34.

Rosenblueth E. (1991) Here and Henceforth, in *Sixth International Conference on Applications of Statistics and Probability in Civil Engineering* (L. Esteva and S.E.Ruiz, eds.), Vol. 3, Mexico City, 81–94.

Rzhanitsyn A.R. (1947) Determination of Safety Factor in Construction, *Stroitel'naya Promishlennost*, (8) (in Russian).

Rzhanitsyn A. R. (1954) *Design of Construction with Materials' Plastic Properties Taken into Account*, Gosudarstvennoe Izdatel'stvo Literatury Po Stroitel'stvu i Arkhitekture, Moscow, 1949 (First edition) 1954, (second edition), Chapter 14 (in Russian) (see also a French translation: A. R. Rjanitsyn: *Calcul à la rupture et plasticité des constructions*, Eyrolles, Paris, 1959).

Rzhanitsyn A.R. (1957) It is Necessary to Improve the Norms of Analysis of Building Constructions? *Stroitelnaya Promyshlennost*, (8), 1957 (in Russian).

Rzhanitsyn A.R. (1971) Determination of the Safety at Loads, Representing Random Processes, *StroitelnayaMekhanika i RaschetSooruzhenii*, (3), 7–11 (in Russian).

Rzhanitsyn A.R. (1973) Economic Principle of Design for Safety, *StroitelnayaMekhanika i RaschetSooruzhenii*, (3), 3–5 (in Russian).Rzhanitsyn A.R. (1978) *Theory of Structural Reliability Analysis in Design*, Moscow: Stroyizdat Publishing House (in Russian).

Sacks J. and Ylvisaker D. (1972) A Note on Huber's Robust Estimation of a Location Parameter, *The Annals of Mathematical Statistics*, Vol. 43 (4), 1068–1075.

Saltelli A., Ratto M., Andres T., CampolongoF., Cariboni J., Gatelli D., Saisana M., and Tarantola S. (2008) *Global Sensitivity Analysis: The Primer*, New York: Wiley.

Sarton G. (1952) *A Guide to the History of Science* (Horus), Waltham: Chronica Botanica Co.

Savoia M. (2002) Structural Reliability Analysis through Fuzzy Number Approach, with Application to Stability, *Computers & Structures*, Vol. 80 (12), 1087–1102.

Savoia M., Ferracuti B. and Elishakoff I. (2005) Fuzzy Safety Factor, in *Safety and Reliability of Engineering Systems and Structures* (G. Augusti, G.I. SchuëllerandM. Ciampoli, eds.), Rotterdam: Millpress, 1783–1791.

Spanos P. D. and Zeldin B.A. (1998) Monte Carlo Treatment of Random Fields: Abroad Perspective, *Applied Mechanics Reviews*, Vol. 51, 219–237.

Scherman N. (ed.) (2001) *Tanakh, Vol. I, The Torah* (Stone edition), New York: Mesorah Publications, Ltd.

Scherer R.J. and Schuëller, G.I. (1998) Quantification of Local Site and Regional Effects on the Nonstationarity of Earthquake Accelerations, in 9th World Conf. on Earthquake Engineering.

Schuëller G.I. (1981) *Einführung in die Sicherheit und Zuverlässigkeit von Tragwerken*, Berlin: Verlag von Wilhelm Ernst & Sohn (in German).

Schuëller G.I. (1987) A Prospective Study of Materials Based on Stockastic Methods, *Material and Structures – Matériaux et Contructions*, Vol. 20, 242–247.

Schuëller G.I. (2000) Recent Developments in Structural Computational Stochastic Mechanics, in *Computational Mechanics for the Twenty First Century* (B.H.V. Topping, ed.), Edinburgh: Saxe-Coburg Publishers, 281–310.

Schuëller G.I. and Ang A.H-S. (1992) Advances in Structural Reliability, *Nuclear Engineering and Design*, Vol. 134, 121–140.

Schuëller G.I and Freudenthal A.M. (1972) Scatter Factor and Reliability of Structures, *NASA CR-2100*.

Schuëller G.I. and Kafka P. (1978) Future of Structural Reliability Methodology in Nuclear Power Plant Technology, *Nuclear Engineering and Design*, Vol. 50, 201–205.

Schuëller G.I. and Strix R. (1987) A Critical Appraisal of Methods to Determine Failure Probabilities, *Journal of Structural Safety*, Vol. 4, 293–309.

Shestov L. (2018) *All Things are Possible*, Grand Rapids, MI: Credo Four Publishing.

Shigley J.E. (1977) *Mechanical Engineering Design* (third edition), New York: McGraw Hill.

Shinozuka M. and Deodatis G. (1991) Simulation of Stochastic Processes by Spectral Representation, *Applied Mechanics Reviews*, Vol. 44, 191–203.

Shinozuka M. and Deodatis G. (1996) Simulation of Multi-Dimensional Gaussian Stochastic Fields by Spectral Representation, *Applied Mechanics Reviews*, Vol. 49, 29–53.

Singer J., Abramovich H. and Yaffe R. (1978) Initial Imperfection Measurements of Integrally Stiffened Cylindrical Shells, TAE Report 330, Technion–I.I.T.

Sinitsyn A.P. (1985) *Design of Structures on the Base of Risk Theory*, Moscow: "Stroyizdat" Publishing House (in Russian).

Smith G.N. (1986) *Probability and Statistics in Civil Engineering*, London: Collins Professional and Technical Books.

Snarškis B.I. (1962) Statistical-Economical Based Safety Margins of Bearing Capacity of Structures, *Proceedings, Lithuanian Academy of Science*, Ser.B, Issue 2(29) and 3 (1963) Vilni Issue and us (in Russian).

Snarskis B.I. (1972) Connection between Methods of Optimal Design Values with Limit State Method, *Reliability Problems in Structural Design,* Sverdlovsk, 206–211 (in Russian).

Son Y.K. and Savage G. J. (2007) Optimal Probabilistic Design of the Dynamic Performance of a Vibration Absorber, *Journal of Sound and Vibration*, Vol. 307 (1–2), 20–37.

Spaethe G. (1987) *Die Sicherheit Tragender Baukonstrutionen*, Berlin: VEB (in German).

Srinivasan S.K. and Vasudevan R. (1971) *Introduction to Random Differential Equations and Their Applications*, Amsterdam: Elsevier.

ST 2394 (2015) General Principles on Reliability for Structures, ISO TK98.

Stevenson J. and Moses F. (1970) Reliability Analysis of Frame Structures, *Journal of Structural Division*, Vol. 96 (11), 2409–2427.

Streletsky N.S. (1840) About a Question of Determining the Allowable Stresses, *Stroitelnaya Promyshlennost*, No. 7 (in Russian).

Streletsky N.S. (1935) Towards the Analysis of the General Safety Factor, *Proekt i Standart*, No. 10 (in Russian).

Streletsky N.S. (1936) On the Problem of General Safety Factor, *Journal of Design & Standard*, #10 (in Russian).

Streletsky N.S. (1937) The Factor of Safety as an Indicator of Strength Equality of Structures, *Comptes Rendus Doklady de L'Academie des Sciences de L'URSS*, Vol. XIV (8), 487–489.

Streletsky N.S. (1940) About a Possibility of Increasing the Allowable Stresses, *Stroitelnaya Promyshlennost*, No. 2–3 (in Russian).

Streletsky N.S. (1946) Practical Consequences of Statistical Analysis of Safety Factor, *Stroitelnaya Promishlenmost*, No. 10, 5–8 (in Russian).

Streletsky N.S. (1947) About Establishment of Safety Factors of Structures, *Izvestiya Akademii Nauk SSSR, Otdelenie Tekhnicheskikh Nauk*, No. 1, 15–26 (in Russian).

Streletsky N. S. (1947) *Statistical Basis of the Safety Factor of Structures*, Moscow: "Stroyizdat" Publishing House (in Russian).

Streletsky N.S. (1963) Modern State of the Design Problem of Construction, *Izvestiya Visshikh Uchebnykh Zavedenii, Stroitelstvo I Arkhitektura*, No. 1, 3–11 (in Russian).

Streletsky N.S. (1975) *Izbrannye Trudy* (*Selected Works*) (Belenia E.I., ed.), Moscow: "Stroiizdat" Publishers (in Russian).

Su H.L. (1959) Statistical Approach to Structural Design, *Proceedings of the Institution of Civil Engineers*, Vol. 13, 353–362.

Su H.L. (1961) Philosophical Aspects of Structural Design, *Journal of Structural Division*, Vol. 87 (5), 1–16.

Su H.L. (1962) Closure to "Philosophical Aspects of Structural Design". *Journal of the Structural Division*, Vol. 88 (4), 209–210.

Sveshnikov A.A. (1968) *Applied Methods Theory of Random Functions*, Moscow: "Nauka" Publishing House (in Russian).

Takewaki I. (2007) *Critical Excitation Methods in Earthquake Engineering*, Amsterdam: Elsevier.

Takewaki I., Abbas M., and Fujita K. (2013) *Improving the Earthquake Resilience of Buildings: The Worst-Case Approach*, Berlin: Springer.

Takewaki I. and Kojima K. (2021) *An Impulse and Earthquake Energy balance Approach in Structural Dynamics*, Boca Raton: CRC Press.

Tankard J. (1979) The H.G. Wells Quote on Statistics: A Question of Accuracy, *Historia Mathematica*, Vol. 6 (1), 30–33.

Thoft-Christensen P. and Baker M.J. (1982) *Structural Reliability: Theory and Applications*, Berlin: Springer.

Thomas L. (1980) On Science and Certainty, *Discover Magazine*, October, 58.

Tikhonov V.I. (1970) *Outliers of Stochastic Processes*, Moscow: "Nauka" Publishing House (in Russian).

Tilahun S.L. and Ngnotchouye J.N. T. (2018) *Optimization Techniques for Problem Solving in Uncertainty*, Hersheley, PA: IGI Global.

Timoshenko S.P. and Gere J.M. (1963) *Theory of Elastic Stability*, New York: McGraw-Hill.Tsikerman L.Y. (1977) *Computerized Diagnostics of Pipeline Corrosion*, Moscow: "Nedra" PublishingHouse (in Russian).

Turkstra C.J. (1970) *Probabilistic Design Formats Theory of Structural Design Decisions, Solid Mechanics Division*, University of Waterloo, SM Studies, Series, No. 2.

Uritckiy M.R. (1973) Homogeneity of the Mechanical Properties of Small Carbonaceous Steel in the Butch of Sheeted Rolling (in Russian). *J. Industrial Construction*, Vol. 12, 25–26.

Ushakov I. (2013) *Optimal Resource Allocation (with Practical Statistical Applications and Theory)*, Hoboken, NJ: Wiley.

Vanmarcke E.H. (1979) Some Recent Developments in Random Vibration, Applied Mechanics Reviews, Vol. 32 (10), 1197–1202.

Vasile M. (ed.) (2020) *Optimization under Uncertainty with Applications to Aerospace Engineering*, Cham: Springer.

Veneziano D. (1978) *Random Processes for Engineering Applications* (Lecture Notes), Department of Civil Engineering, M.I.T., Cambridge, MA.

Volmir A.C. (1967) *Stability of Deformable Systems*, Moscow: "Nauka" Publishing House (in Russian).

Wasfy T.M. and Noor A.K. (1998) Finite Element Analysis of Flexible Multibody Systems with Fuzzy Parameters, *Computer Methods in Applied Mechanics and Engineering*, Vol. 160 (3–4), 223–243.

Wells H.G. (1903) *Mankind in the Making* (republication: Frankfurt am Main: Outlook, 2018).

Wen Y.K. (1977) Statistical Combination of Extreme Loads, *Journal of the Structural Division, Proc. ASCE* 103, ST5 1079.

Wen Y.K. (1993) Reliability-Based Design under Multiple Loads, *Structural Safety*, Vol. 13 (1–2), 3–19.

Wentzel E.S. (1972) *Study of Operations*, Moscow: Soviet Radio (in Russian).

Wierzbicki W. (1936) *Safety of Structures as a Probability Problem*, Warsaw: Przeglad Techniczny.

Wikipedia, Blind Men, and an Elephant, available at https://en.wikipedia.org/wiki/Blind_men_and_an_elephant#:~:text=The%20tale%20later%20became%20well,no%20one%20has%20fully%20experienced., accessed on May 27, 2021.

Wilks S.S. (1951) Presidential address to the American Statistical Association, *Journal of American Statistical Association*, Vol. 46 (253), 1–18.

Xie M. and Shen K. (1989) On Ranking of System Components with Respect to Different Improvement Action, *Microelectronics Reliability*, Vol. 29, 159–164.

Yao J.M.T. (1994) Reliability Evaluation of a Complex system, in *Risk Analysis* (A.S. Nowak, ed.), Ann Arbor: University of Michigan, 249–252.

Zadeh L.A. (1998) Some Reflections on Soft Computing, Granular Computing and Their Roles in the Conception, Design and Utilization of Information/Intelligent Systems, *Soft Computing*, Vol.2 (1), 23–25.

Zadeh L.A. (2004) *Fuzzy Logic Systems, Origin, Concepts and Trends*, available at http://wi-consortium.org/wicweb/pdf/Zadeh.pdf, November 10, 2004 (accessed on July 11, 2021).

Zadeh L.A. (2011) My Life and Work – A Retrospective, *Applied Computer Mathematics*, Vol. 10 (1), Special Issue, 4–10 (available at www.cs.berkeley.edu/~zadeh/papers/Preface%20ACM-My%20Life%20and%20Work--A%20Retrospective%20View.pdf (Source: https://quotepark.com/authors/rudolf-e-kalman/, accessed on July 11, 2021).

Zadeh L.A. (2015), available at https://libquotes.com/lotfi-a-zadeh (accessed on July 11, 2021).

Zarenin Y.G and Zbirko M.D. (1971) On Reliability of Systems with Protection, *Technical Cybernetics*, (2), 46–52 (in Russian).

Zaslavsky G. M. (1984) *Stochasticity of Dynamic Systems*, Moscow: "Nauka" Publishing House (in Russian).

Zhao R. and Goving R. (1991) Defuzzification of Fuzzy Intervals, *Fuzzy Sets and Systems*, Vol. 43, 45–55.

Author Index

Subject Index

Note: Page numbers in *italics* indicate figures and in **bold** indicate tables on the corresponding pages.